E17.

DR KARL
KRUSZELNICKI

HarperCollins*Publishers*

HarperCollins*Publishers*

First published in Australia in 2006
by HarperCollins*Publishers* Australia Pty Limited
ABN 36 009 913 517
www.harpercollins.com.au

HarperCollins*Publishers*
25 Ryde Road, Pymble, Sydney, NSW 2073, Australia
31 View Road, Glenfield, Auckland 10, New Zealand
77–85 Fulham Palace Road, London W6 8JB, United Kingdom
2 Bloor Street East, 20th floor,Toronto, Ontario M4W 1A8, Canada
10 East 53rd Street, New York, NY 10022, USA

National Library of Australia Cataloguing-in-Publication data:
Kruszelnicki, Karl, 1948– .
It ain't necessarily so ... bro.
Bibliography.
ISBN 13: 978 07322 8061 1.
ISBN 10: 0 7322 8061 3.
1. Science – Popular works. 2. Discoveries in science.
3. Curiosities and wonders. I. Title.
500

Cover photograph by Gerald Diel
Cover design by Adam Yazxhi, Maxco Creative
Illustrations for 'Daddy long legs', 'Boiling frogs', 'Four leaf clover', 'Dogs and chocolate' contributed to by Max Yazxhi (aged 3½ years)
Internal design and layout by Judi Rowe, Agave Creative Design
Printed and bound in Australia by Griffin Press.

79gsm Bulky Paperback used by HarperCollins*Publishers* is a natural, recyclable product made from wood grown in a combination of sustainable plantation and regrowth forests. It also contains up to a 20% portion of recycled fibre.The manufacturing processes conform to the environmental regulations in Tasmania, the place of manufacture.

10 9 8 7 6 07 08 09 10

This book is dedicated to Pluto,
a small compensation for your loss of status ...
In 1930, a farm boy discovered Pluto, which was
in the right place, but for the wrong reasons.

CONTENTS

YOUNG EARTH

We humans have a deep sense of curiosity and want to understand the world around us. One question that we often ask is, 'How old is our world?'. With today's knowledge, science tells us that the Earth is about 4.6 billion years old, but a small number of people — the 'Young Earthers' — stick ferociously to a belief that our planet is 6000 years old.

History of Earth's Age

Back around 400 AD, St Jerome, the Italian scholar and priest, made a 6000-year estimate for the age of the Earth, as did the later scholars, Scaliger and Venerable Bede. Around 1600, in the Shakespearean play *As You Like It* (Act IV, Scene 1), Rosalind says, 'The poor world is almost six thousand years old …'

In the Bible, Psalm 90, Verse 4 reads, 'For a thousand years in your sight are like a day that has just gone by, or like a watch in the night'. In other words, the six Days of Creation could take 6000 years to pass, which would fit in well with the Earth being roughly 6000 years old.

In 1642, Dr John Lightfoot, English minister, rabbinical and linguistic scholar and later Vice Chancellor of Cambridge University in the United Kingdom, wrote his *Observations on Genesis* — a book of some 20 pages. In it, he observed that Man (not the World) was created at 9.00 am. He based this on *Genesis*, Verse 26, Chapter 1: 'Man was created by the Trinity about the third hour of the day, or nine of the clock in the morning.' Two years later, on the basis of further interpretation of the Bible, he wrote that the Earth was created on Sunday, 12 September 3928 BC. He also estimated that Man was created five days later, on Friday, 17 September.

Ussher, Mega-brain

The 'Young Earthers' use the planet-dating estimate which came from James Ussher (1581–1665), a gifted linguist and prolific religious scholar.

Ussher entered Trinity College in Dublin at the age of 13, received a master's degree at the age of 20, was ordained as a priest at 21 and appointed Professor of Theology at Trinity when he was only 26. He was also twice Vice Chancellor of Trinity, in 1614 and 1617. In 1625, he was appointed Archbishop of Amargh, and by 1634 was Primate of All Ireland. A prolific writer, he published some 17 volumes. Ussher was a very smart and hard-working dude.

In 1650, he published the first part of *Annals of the Old Testament, Deduced from the First Origins of the World* in Latin. The second part was published four years later. It was an immense work, covering everything from the creation of the world to the dispersion of the Jews in the reign of Vespasian (69–79 AD). Fuller translated the *Annals* into English in 1658.

How Ussher Did It

Archbishop Ussher used the best of his considerable historical and scholastic skills to deal with the poor historical and archeological records of the time. He also had to deal with the fact that there were several different versions of the Bible, built up from different sources over several centuries. He studied three different time periods.

The first time period — Early Times: Creation up to Solomon — was the easiest to calculate. The Bible gives a continuous male lineage from Adam to Solomon — 'Adam begat Seth begat Enosh begat Kenon ...' — together with everybody's ages. All Ussher had

You don't look a day over 4.6 billion years...

The simple question of 'How old is our world?' can bring answers ranging from 6000 to 4.6 billion years. Let's go to the adjudicators ...

to do was add the ages together. But there was one minor problem — different versions of the Bible gave different ages. So Ussher simply used the Hebrew Bible.

The second time period — Early Age of Kings: Solomon to the Destruction of the Temple and the Babylonian Captivity — was more difficult. Once the 'begats' ran out, Ussher had only the lengths of kings' reigns in the Bible to work with. So he cross-referenced these with the then known historical records to continue his timeline.

In the third time period — Late Age of Kings: Ezra and Nehemiah up to the Birth of Jesus — Ussher had to link individual events in the Bible with the historical records of the relevant societies. For example, 2 *Kings* 25:27 reads: 'In the thirty-seventh year of the exile of Jehoiachin, king of Judah, in the year Evil-merodach became King of Babylon, he released Jehoiachin from prison on the twenty-seventh day of the twelfth month.' From historical non-biblical records, this can be separately related to Nebuchadnezzar II's death.

Ussher Into Bible

This hard work gave Ussher a date of about 4000 BC for the Creation. The scholars of the day already knew of the counting error that Dionysius Exiguus (Dennis the Small), who set up the AD year counting system, had accidentally made. Thanks to this error, Jesus Christ was probably born in 4 BC. So Ussher added four years to 4000 BC to give 4004 BC as the date when God created Life, the Universe and Everything.

In his *Annals*, Ussher stated that the world was created on Sunday, 23 October 4004 BC.

The 'Sunday' was easy to calculate. The Bible tells us that God rested on the seventh day, which, under the Hebrew system, was Saturday. Counting backwards to the first day gives us Sunday.

He came up with '23 October' by using the slightly incorrect astronomical tables of the day to find the autumnal equinox (when the hours of daylight and darkness are equal). He chose the autumnal equinox because it was the beginning of Jewish calendar year.

In 1701, this estimate for the world's creation found favour with Bishop William Lloyd of Winchester, who got the publishers, Clarendon Press at Oxford, to insert it in the Great Edition of the King James Bible. They placed his dates — without authorisation! — in the margins of the appropriate pages of *Genesis*, where they remained for centuries. There are no footnotes or explanations in the Bible to justify how these dates came to be inserted.

The fact that this unauthorised date appeared in the margins of the Bible made it 'gospel truth' for most Christians back then, and for a much smaller percentage today.

'Young Earthers'

Today's 'Young Earthers' — a very small but passionate creationist group — still use Ussher's estimate of 4004 BC. (Their interpretation is at odds with most Christian faiths, including the Catholic Church and the Church of England.) They get around the geological and other scientific evidence with extreme mental gymnastics, by suspending many of the known laws of science.

For example, they claim that in the past the continents drifted at kilometres per hour rather than centimetres per year. They also claim that coral reefs formed at 40 000 times their present rate, that oceans evaporated at 4 m per day to form salt beds, that ocean floor sediments formed at 80 million times their present rate, and so on.

Big It Up for Ussher

Today, many people mock James Ussher. But evolutionist Stephen Jay Gould, who disagreed with Ussher's 6000-year estimate, nevertheless respected him as a scholar. He wrote: 'I shall be defending Ussher's chronology as an honourable effort for its time, and arguing that our usual ridicule only records a lamentable small-mindedness based on mistaken use of present criteria to judge a distant and different past.'

It is my own personal belief that if Archbishop Ussher were alive today, he would look at the evidence with his keen mind, and happily accept the current 4.6-billion-year estimate of the age of the Earth …

Other Religions

Zoraster, a Persian prophet from the 6th century BC, set the age of the Earth at 12 000 years. The priesthood of Chaldea in ancient Babylonia set it much older — at two million years. The Brahmians of India took it to the max — they saw both the Earth and Time as eternal. Surprisingly, and true to their name, the 'Young Earthers' estimate of 6000 years is probably the youngest figure on record.

Science Enters

James Hutton, often known as the 'Father of Modern Geology', brought science to the discussion of the age of the Earth. In 1785, he read his essay, 'Theory of the Earth' to the Royal Society of Edinburgh. He discussed how the land that his audience walked on had been made by the rivers and seas of past ages, and how the time needed to lay down this land was truly immense. His early critics scoffed at him for 'running about the hillsides with a hammer to find how the world was made', and it was decades before his theories of geology were accepted. He knew that the time needed to lay down the land was immense, but he didn't know how long.

Charles Lyell, born in 1797, the year that Hutton died, continued Hutton's work. Lyell went to the volcano of Etna on Sicily, and realised that each new layer of molten lava was deposited on the layer beneath, and so on, and so on. He knew the height of the volcano, how quickly it grew and how often it erupted. Simple maths gave him an age for the volcano in the hundreds of thousands of years.

Theological Backlash

Ussher was a clever man. He used the best historical, biblical and astronomical data of his day to try to work out the age of the Earth. But in later times he fell out of favour, even with theologians.

In 1890, Dr William Henry Green, Professor of Old Testament at the Princeton Theological Seminary, re-analysed Ussher's work in a more modern context and demolished it in his paper, *Primeval Chronology*. He concluded, '... The Scriptures furnish no data for a chronological computation prior to the life of Abraham; and the Mosaic records do not fix, and were not intended to fix, the precise date either of the Flood, or the Creation of the World.'

References

Brice, W.R., 'Bishop Ussher, John Lightfoot and the age of creation', *Journal of Geological Education*, 1982, Vol 30, pp 18-24.

Gould, S.J., 'Fall in the house of Ussher', in S.J. Gould, *Eight Little Piggies: Reflections in Natural History*, New York: Norton, 1993, p 183.

Lewis, Cherry, *The Dating Game*, Cambridge: Cambridge University Press, 2000, pp 12-26.

Wise, Donald U., 'Creationism's geologic time scale', *American Scientist*, 1998, Vol 86, pp 160-173.

MYSTERIOUS KILLER CHEMICAL

Early in 2006, commercial fishers were forbidden to ply their trade in Sydney Harbour. The problem was toxic quantities of a nasty chemical, dioxin, getting into their fish from sediment on the harbour floor. The problem had been present for many years but had been ignored. Unfortunately, this happens with many nasty chemicals. The elected officials simply hope that the expensive cleanup will be left for the next government in power.

Chemical DHMO

Consider the chemical DiHydrogen MonOxide, usually called DHMO.

It is found in many different cancers, but there is no proven causal link between its presence and the cancers in which it lurks — so far. The figures are astonishingly high — DHMO has been found in over 95% of all fatal cervical cancers, and in over 85% of all cancers collected from terminal cancer patients. Surprisingly, some elite athletes will load up with DHMO before they participate in endurance sports such as cycling and running. However, the athletes later find that withdrawal from DHMO can be difficult, and

DHMO ... a chemical MOFO

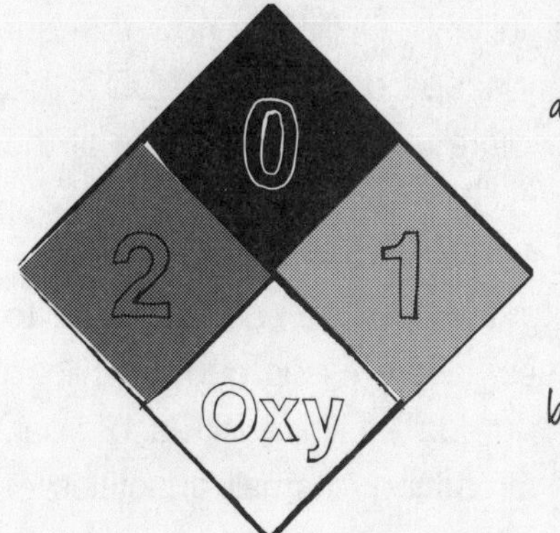

DHMO is still widely used as an industrial solvent and coolent, and as a fire retardant and suppressant. It is also essential in the manufacture of biological and chemical weapons, and in nuclear power plants.

DiHydrogen MonOxide
aka Hydric Acid,
Hydronium Hydroxide
and ... Water

even fatal. Medically, it is almost always involved in diseases that have sweating, vomiting and diarrhoea as their symptoms.

Despite these known associations, DHMO is still widely used as an industrial solvent and coolant, and as a fire retardant and suppressant. It is essential in the manufacture of biological and chemical weapons — and in nuclear power plants. While it has many industrial uses, it is cheap enough to be dumped casually into the environment, where it has many unwanted side effects. DHMO is a major contributor to acid rain and is heavily involved in the Greenhouse Effect. In industry, it can short out electrical circuits. It causes corrosion of some metals and can reduce the efficiency of your car's brakes.

It is used to help distribute pesticides and herbicides that have known side effects. In fact, no matter how well you wash your fruits and vegetables, trace amounts of DHMO will always remain. And in the environment where the fruits and vegetables were grown, long after the pesticides and herbicides have degraded away, the DHMO will remain, because it is so stable. Indeed, DHMO is now thought to be a significant contributor to landscape erosion. Like DDT, this chemical has been found in the desolate remote wastelands of the Antarctic.

One reason why DHMO can be so dangerous is its chameleon-like ability to not only infiltrate into the background but also to change its state. As a solid, it causes severe tissue burns, while in its hot gaseous state it kills hundreds of people each year. Thousands more die each year by breathing in small quantities of liquid DHMO into their lungs.

The Bans

In 1990, at the Santa Cruz campus of the University of California, Eric Lechner and Lars Norpchen publicised the dangers of DHMO — DiHydrogen MonOxide. Enough people had begun to use the Internet by 1994 to give Craig Jackson an ideal forum (via his web page) to set up *The Coalition to Ban DHMO*. Slowly, awareness of this chemical spread. In 1997, 14-year-old Nathan Zohner at the Eagle Rock Junior High School in Idaho told 50 of his fellow students about DHMO. He then surveyed their attitudes — and 43 of them signed a petition to ban this chemical immediately.

In March 2004, the small city of Aliso Viejo in Orange County in California began a process to ban DHMO. An enthusiastic paralegal on the Aliso Viejo city payroll had read of DHMO's evil properties on the Internet, specifically its use in the production of styrofoam containers. As a direct result, a motion to ban styrofoam containers was placed on the official agenda of the next meeting of the council.

Luckily for the reputation of the city, the motion was withdrawn before it could be voted on.

Why luckily, you ask?

Da Dahhhh!

Well, DHMO, DiHydrogen MonOxide — also known as Hydric Acid and Hydronium Hydroxide — is usually called just plain water. First-year university chemistry students have made laboured jokes about water for years.

But, here's the point about misinformation, or disinformation.

You can give people this totally accurate (but emotionally laden and sensationalist) information about water and then, when you survey them, 70–90% will willingly sign a petition to ban it. And it doesn't matter where in the world you do the survey.

In the case of Nathan Zohner, his 50 fellow year nine students were studying science. Many of them had parents who worked in the nearby Idaho Nuclear Engineering and Environmental Laboratory. The students could have asked their science teacher for advice — but none did. Of the students, 43 signed the petition to ban water, six were undecided, while one recognised DHMO for what it was and would not sign.

We live under the illusion that we understand the world around us, but in reality, very few of us can change a car's spark plug, or the memory or hard drive in our computer. In 1997, Nathan Zohner from Eagle Rock, Idaho, won a Science Fair Prize for his project. It was called, 'How Gullible Are We?'.

Perhaps the answer is, 'pretty gullible', depending on our particular field of ignorance.

World's Worst Carcinogen

Citric acid has been described as the most dangerous carcinogen known to the human race — according to a letter that supposedly originated from a Paris hospital in 1974. It listed 139 dangerous food additives, with first position reserved for citric acid.

Arnold Bender, Professor of Nutrition and Dietetics at the University of London, debunked this hoax. Even so, in July 1976, the French Minister of Agriculture had to explain to the French Senate that citric acid was perfectly harmless, and that his department was trying to find out who started the hoax. But this list resurfaces every few years in various countries around the world.

Citric acid occurs naturally in many fruits, e.g. tomatoes, pineapples, strawberries, oranges and lemons. It is also used as a food additive (additive code 330) because its natural tartness balances beautifully against the sweetness of sugar. As an extra advantage, it's an antioxidant that works with other antioxidants to preserve foods and stop them from spoiling. When it is used as a food additive, the amount added is usually less than what is already naturally present in many foods.

However, there's another very good reason why we should not worry about citric acid — our bodies make it all the time. When I was a first-year medical student, a large part of the biochemistry lectures was devoted to the Citric Acid Cycle, also called the TCA (Tricarboxylic Acid) Cycle or the Krebs Cycle. Sir Hans Krebs won the Nobel Prize in 1953 for his work in describing this cycle. The cycle is absolutely essential to our metabolism when we convert glucose to energy. As part of this cycle, citric acid is manufactured.

Citric acid is made millions of times every day in every cell of the body — and yet, somebody tried to convince us that it was the most dangerous carcinogen in the known Universe.

References

Butler, Kathy, 'Science, sex and The Simpsons', *The Skeptic* (Australia), Winter 2000, Vol 20, No 2, p 5.

Cardwell, Glenn, 'The hoax is on us', *The Skeptic* (Australia), Winter 1998, Vol 18, No 2, pp 37, 38.

Encyclopaedia Britannica, Ultimate Reference Suite DVD, 2006 (accessed 14 May 2006) — 'tricarboxylic acid cycle'.

Gorard, Stephen, 'Fostering scepticism: the importance of warranting claims', *Evaluation and Research in Education*, 2002, Vol 16, No 3, pp 136-148.

http://www.hmo.org

COMA AND THE TV SOAP OPERA

The TV soap opera is a sadly underrated art form. I myself was blessed with a most prestigious honour — yes, I actually appeared on *Neighbours*.

In Episode 4550 (which went to air in Science Week, August 2004), I, a real-life Dr Karl, played opposite an on-screen, in-house *Neighbours* character, also called Dr Karl. We had a curiously existentialist dialogue where we subtly explored the boundaries of personal separation and ego, while superficially appearing to be discussing a mere case of mistaken identity ('What, *you're* Dr Karl? So am I.'). However, as much as I do love my TV soaps, I have to admit that they sometimes bend the truth — excluding *Neighbours*, of course. A good example is the TV coma.

Coma

Medically speaking, a coma is a state of unconsciousness, in which the patient does not show any spontaneous activity and does not respond to external stimuli.

There are many causes of coma, but they can be broadly broken into two categories. The first category includes traumatic causes such as head injury, smoke inhalation, motor vehicle collisions and falls, and (God bless the soaps) shipwrecks and aeroplane crashes. The second category involves non-traumatic causes, e.g. poisoning, stroke, heart attack and diabetes.

Either way, some coma patients die, some recover full function and some stay deeply unconscious. There are also cases where coma patients are unaware of themselves and the environment and yet they show sleep–wake cycles. (This condition is known as a 'persistent vegetative state'.)

The eyes of a person in a coma can be open. The arm and leg muscles after a while contract (and bend the limbs), but also lose strength rapidly and waste away. Many coma patients need to be given air and food through tubes. They also suffer skin ulcers, as well as bladder and bowel incontinence.

Finding TV Comas

In December 2005, Dr David Casarett and his colleagues published their ground-breaking paper, 'Epidemiology and prognosis of coma in daytime television dramas', in the *British Medical Journal* (BMJ). They had examined people's perceptions of, and attitudes to, comas.

They trawled through a decade of TV soaps (1995–2005), including *Days of Our Lives, The Young and the Restless, The Bold and the Beautiful, General Hospital* and *Passions*. They found a total of 73 comas, but had to disqualify nine of them for various reasons. One character woke up for meals, two were faking it, three seemed to be drug-induced, and there were a final three in whom the writers seemed to lose interest (so the audience never found out what happened to them). This left 64 people in comas.

TV Coma Mistakes

There were many mistakes.

The most common was the 'Sleeping Beauty' syndrome, where the coma patients slept with their eyes shut, were well tanned and well groomed and had normal muscle tone.

There's no telling if and when he will recover...

TV soaps sometimes bend the truth ... the way a coma is shown on TV differs greatly from what happens in real life.

But let's look at the dramatic stuff, that TV loves to deal with.

First, how many died? With the TV coma patients who had suffered trauma, the death rate was 6% (vs 67% in real life). For non-trauma coma patients, the death rate was 4% (vs 53% in real life). By the way, in the TV soaps, two of the people who 'died' were later revealed to be actually alive (but one was replaced by a mannequin). The authors 'counted them as deaths, because we reasoned that viewers would perceive them as having died' — and the study was about viewer perceptions of coma.

Second, most people who do survive a coma suffer disabilities. But in TV-land, 89% of trauma coma patients pull through and return to full function with no disabilities (vs 7% in real life). For non-trauma coma patients, 91% of TV coma victims pull through (vs 1% in real life). Furthermore, in the real world, the vast majority of coma patients need months of rehabilitation, while in TV soaps, 86% awaken, hop out of bed and immediately kiss their secret lover.

The TV coma victims did however, suffer curious personality changes and memory lapses, which helped push the plot in unexpected directions.

Deep Soap

Unexpected directions are why TV soaps have such loyal audiences. The characters are deep and complex, and the many interweaving stories form a tangle of intrigue and subplots.

Soap operas attract huge audiences. *Coronation Street* attracted 13 million viewers per episode in 2001, while in the USA, soapie audiences number around 40 million people — and American soaps are popular in 90 countries.

TV soaps exist to entertain the audience and keep them glued to the telly to help sell the advertisers' products. So, as a rule, the writers don't want them to be too depressing. Oddly, however, another study in the *BMJ* showed that death rates in British soapies are up to seven times higher than in the general population. So the people in *EastEnders* lead more perilous lives than Formula One drivers, oil rig divers and bomb disposal experts.

Perhaps they have to compensate for the very high death rates by making comas so survivable …

The Power of Soap

TV soap operas have the power to change people's beliefs and actions.

In April/May 2001, Alma in the famous British soapie *Coronation Street* developed cervical cancer. In the United Kingdom there was an immediate 20% increase in the number of pap smears (which can diagnose cervical cancer) performed. This lasted until mid-August (six weeks after the character died).

Coma and Movies

As you might have guessed, movies are just as bad as TV soap operas in how they depict comas.

Dr Wijdicks, a neurologist, found 30 movies made between 1970 and 2004 that depict a character in a prolonged coma. He showed 22 clips from 17 of these movies to a viewing audience of educated, mature-age people. Only two of these movies — *Dream Life of Angels* and *Reversal of Fortune* — were reasonably accurate.

In general, the movies were lax in showing the great complexity of care needed to keep a comatose patient alive. For example, one patient had only a simple nasal air tube, but the soundtrack was that of a respirator, which would involve a large tube down the throat. They were also usually wrong about the cause of the coma and the probable chance of the patient awakening. Finally, the movies did not show an appropriate level of compassionate discussion between the medical team and the family members.

The viewing audience could not pick the errors in one-third of cases. They were asked what they would do if they suddenly found themselves in the situation of having to deal with a family member or friend in a prolonged coma. Predictably, in 40% of cases, they said they would use what they had learnt about comas from the movies — even though it was wrong.

What's so Good About Truth?

The movie critic, Roger Ebert, knows what he wants in a movie.

He wrote: 'I want moods, tones, fears, imaginings, whims, speculations, nightmares. As a general principle, I believe films are the wrong medium for fact. Fact belongs in print. Films are about emotions.'

Birth and TV

As you might have guessed, TV soapies also get it wrong with births — not to mention that the newborn babies always look big enough to be two years of age.

First, there is a very high death rate in TV mothers and babies, much higher than in real life. Second, 37% of TV labours are so quick that the babies are delivered unexpectedly, without the medical team, spouse, partner or family members present. Finally, pain relief (e.g. inhalational analgesia, narcotics and epidurals) was given in only 3–7% of TV births.

The lesson is obvious.

If your relative is in a coma, take them to a TV station for an instant cure. But leave right away if there is an imminent birth.

References

Casarett, David, et al., 'Epidemiology and prognosis of coma in daytime television dramas', *British Medical Journal*, 24 December 2005, pp 1537-1539.

Clement, Sarah, 'Television gives a distorted picture of birth as well as death', *British Medical Journal*, 25 July 1998, p 317.

Crayford, Tim, et al., 'Death rates of characters in soap operas on british television: is a government health warning required?', *British Medical Journal*, 20 December 1997, pp 1649-1652.

Howe, Andy, et al., 'The impact of a television soap opera on the NHS Cervical Screening Programme in North West of England', *Journal of Public Health Medicine*, 2002, Vol 24, No 4, pp 294-304.

'Medical Aspects of the Persistent Vegetative State', by The Multi-Society Task Force on PVS, *The New England Journal of Medicine*, 26 May 1994, pp 1499-1508.

Wijdicks, E.F.M., et al., 'The portrayal of coma in contemporary motion pictures', *Neurology*, Masy 2006, pp 1300-1303.

WATER OFF THE DUCT TAPE

A while ago, a tiny spring in the on/off switch on my 20-year-old television set died. 'It's easy,' I said to my wife, 'I'll just pull the 240-volt plug out of the wall, safely discharge the 25 000 volts present on the TV set tube, pull the chassis apart, solder in a new switch and put it all back together again.' Well, she raised her eyebrows at me and said, 'I'll show you "easy",' as she tore a strip of duct tape from a roll that happened to be lying around, and taped the switch into the 'on' position. It was a semi-permanent and effective repair, needing a new strip of duct tape every week or so, a job that takes about 10 seconds each time.

Most people believe that duct tape was invented for air-conditioning ducts — but it wasn't. In fact, for some time, duct tape was not allowed to be used on ducts.

Invention of Duct Tape

Duct tape was invented during World War II for the military, who wanted a strong, waterproof self-adhesive tape to keep water out of ammunition cases. John Denoye and Bill Gross led the team that invented it. They worked at Permacell, a division of Johnson &

Johnson, a company with much experience in producing adhesive surgical tapes. So the team began by experimenting with a surgical tape.

Their first duct tape was a dull greenish, cotton-mesh fabric coated with polyurethane sealant on one side and a rubber-based adhesive on the other. This made it both waterproof and easy to peel off — and very strong. Even better, you didn't need scissors to cut it — you could tear it both longways and crossways, by hand.

Name of Duct Tape

At the time it was called 'duck tape' (as in 'quack, quack'). We don't really know why, but there are three popular theories.

One theory suggests that the soldiers called it 'duck tape', because water rolled off it like off a duck's back. Another theory links the name to the cotton fabric known as 'duck' used in the manufacture of the tape. The third theory claims some kind of link

Duct tape was developed for the military in WWII. They had a need for tape that was strong, waterproof and easy to peel off and as a bonus it could be torn both longways and crossways, by hand.

to an amphibious military vehicle used in World War II. It was called a 'duck', because the manufacturer's code name for it was DUKW.

However, because it was used on ammunition cases, it was also called 'gun tape'. In the racing car business it is called '100-mile-per-hour tape' and '200-mile-per-hour tape', because it will adhere to a car at these speeds. Air force technicians call it '1000-mile-per-hour' tape, because it will adhere to the radome (radar dome) of a jet fighter at this phenomenal speed.

It was called 'duck tape' in a Gimbels department store advertisement of June 1942 ('blinds in cream with cream tape, or in white with duck tape') and in a US government surplus property ad offering '44 108 yards of cotton duck tape' in 1945.

The very first use of the phrase 'duct tape' (with a 'T' as in Tango) seems to be in 1970, when the bankrupt Larry Plotnik Company of Chelsea, Massachusetts, unloaded 14 000 rolls of the stuff.

Perhaps the 'duct' version comes from the Latin word *ducere*, meaning 'to lead' or 'to convey'. It already appears in 'viaduct' (that carries cars or trains over a valley or gorge) and in 'aqueduct' (that carries water over long distances).

Whatever it's called, it's magic stuff.

Banned in Air-Con Ducts!

Thanks to the post-World War II building boom, a variety of duck/duct tape was being used to join sections of air-conditioning ducts. It had now evolved into a silvery version that was stronger, with a more powerful adhesive — and nothing like the stuff you buy in hardware stores. HVAC (Heating, Ventilation and Air Conditioning) professionals wouldn't dream of using the inferior grades of tape.

However, even the top grades were not very good. In the late 1990s, the Lawrence Berkeley National Laboratory (Environmental Technologies Division) in California looked at how much energy escaped from air-con ducts. Surprisingly, they found that clear polyester tape was better than the best HVAC-grade duct tape.

In response, various US government bodies prohibited the use of duct tape in the HVAC industry. So yes, for a while, you weren't allowed to use duct tape on air-con ducts.

Of course, the HVAC industry responded by developing new, improved grades of duct tape.

Other Uses

On 10 February 2003, the US Department of Homeland Security broadcast that a terrorist attack was likely. It advised Americans to buy plastic sheeting and duct tape, in case of a biological or chemical attack. People bought lots of duct tape. In response, the Pressure Sensitive Tape Council announced that its 26 members had 'mobilised to meet the increased demand for duct tape', with the regret that 'we wish we were increasing sales for another reason'. However, Glen Anderson, then Executive Vice President of the Pressure Sensitive Tape Council had mixed feelings about the duct tape itself. He said, 'Polypropylene tape would be most acceptable. Duct tape will leave a residue, whereas polypropylene will not.'

A matte black version of duct tape is used in the theatre, music and entertainment industry, where it is called gaffer tape. It's matte so that it doesn't reflect light. The tape also sticks to things using a special glue that won't leave a residue when you remove the tape. Roadies who do pub gigs tell me that they prefer the Nashua version over the equally fine 3M version — because the Nashua version deals a little better with spilt beer! The film and music industry could not survive without duct tape.

Medicos at the Madigan Army Medical Center in Tacoma, Washington, have found that duct tape can treat warts. Alligator hunters in the bayous of Mississippi and Louisiana use it to strap shut the mouths of alligators. Intertape Polymer Group makes a grade specifically for nuclear reactors. Duct tape is also used in fashion to repair jeans — and boost cleavage.

Do Anything Tape ...

A whole mythology has risen around duct tape.

In Scandinavia, it's called 'Jesus tape', because it can fix anything. Some people, including John Leland of *The New York Times*, think of duct tape as 'national shorthand for a job done almost right'. In other words, it could be used to fix things, usually temporarily, usually by unskilled people, and usually with the bare minimum of time and effort. This might be true in some cases, but not always.

In April 1970, the astronauts of the crippled *Apollo 13* mission used duct tape to make an emergency carbon dioxide scrubber, which kept them alive. An explosion had crippled the spacecraft on its way to the moon, just 55 hours after takeoff. Ed Smylie, one of the NASA engineers on the ground, designed the life-saving cardboard/duct tape assembly, using only material available to the astronauts in the spaceship. For nearly four days the three astronauts had to remain in the lunar module, which was designed to keep two men alive for only two days. One problem that had to be solved was how to remove the carbon dioxide from the limited air supply. Ed Smylie said that he wasn't worried, once he knew that the astronauts had duct tape on board. In 2005 he said, 'One thing a Southern boy will never say is "I don't think duct tape will fix it".'

It is claimed that the 'do-it-yourselfer' needs only two tools — duct tape to stick stuff together and WD-40 to unstick stuff.

And Carl Zwanzig, the famous sci-fi fan, said, 'Duct tape is like The Force — it has a light side and a dark side, and it binds the Universe together.'

Builder's Tape

A builder from Brisbane emailed me with his preference in duct tape. He called it 'speed tape', because it could strap anything together very quickly.

It was a silvery tape, which had to be cut with a knife. He used it to strap 6 m lengths of dressed pine to his ute. It was quicker to use than ropes or nylon ratchet tie-downs. More importantly, it caused less damage to the sharp edges of the dressed pine, which had to be kept in as good a condition as possible.

Houston Problem Myth

Most people who know the *Apollo 13* story, or who are interested in space travel, know the phrase, 'Houston, we have a problem'.

Close, but not quite ...

Jack Swigert, the command module pilot, said, 'Okay Houston, we've had a problem here.' James A. Lovell, the commander, then said, 'Houston, we've had a problem.'

References

John, Leland, 'Unravelling duct tape, warts and all', *The New York Times*, 16 February 2003.

Safire, William, 'Why a duck?', *The New York Times*, 2 March 2003.

Wohleber, Curt, 'Duct tape', *Invention & Technology*, Summer 2003, pp 12, 13.

BLACKOUT BABY BOOM

We have all heard what happens when a massive, prolonged blackout throws men and women together with no warning – and no television. The story is that they love each very much in a special way and nine months later a miracle happens, and the maternity wards of hospitals are full of happy couples and lovely little reminders of the event. It's a sweet story, but totally untrue.

Power Grid Mathematics

If you have one electrical generating station, you need another one for backup in case the first one dies. If you have five stations, you can still get by with just one for backup. If you have 10 power stations, you don't need a separate backup station because each of the 10 power stations usually has enough reserve capacity (say 10%) to be able to share the extra load between them if one of them stops working. This is one of the big advantages of having electrical power stations tied into a grid.

One of the disadvantages is that the grid becomes very complicated to run.

For example, just to design a simple grid with a few generators means that you *have* to use fancy maths that involve the square root of minus one (the two identical 'imaginary' numbers that when multiplied together give you a negative number). This sounds crazy.

If you multiply two positive numbers together you get a positive number. If you multiply two negative numbers together you get a positive number. It is 'impossible' to multiply a 'real' number (e.g. 2, 12.345, etc.) by itself and get a negative number. Even so, electrical engineers use 'imaginary' numbers to design a grid that feeds power to your fridge, to keep your ice cream cold.

Biggest Blackout Ever

Probably the biggest blackout in history happened on 14 August 2003, affecting 50 million people from Canada to New York and Michigan. It was a week before the power was reliably restored.

The system that crashed was huge. At the time of the blackout, 142 separate regional control rooms oversaw the activities of about 3000 utilities that ran 6000 electrical power plants with a total generating capacity of 61 800 MW.

Unfortunately, the American electrical grid had been deregulated in the 1990s. Cost cutting and profit gouging meant regular maintenance was scrapped. Another factor was the big profits involved in long-distance electrical power sales. This meant that the load on the grid increased, making it more complex to run.

Trees Caused Biggest Blackout

The Final Report of the US–Canada Power System Outage Task Force named four causes for the 2003 blackout.

First was 'inadequate system understanding'. A few of the utilities (FirstEnergy and ECAR) in the area that crashed first did not realise how fragile and vulnerable the overloaded system was.

Second was 'inadequate situational awareness'. In other words, FirstEnergy did not recognise, or understand, that the system was deteriorating and beginning to crash.

Third was 'inadequate tree trimming'. Trees had not been trimmed, because the company wanted to generate bigger short-term profits by cutting back on regular maintenance. The trees grew, as they do, and touched and then shorted out three of FirstEnergy's 345-kV lines and one 138-kV line.

Fourth was 'inadequate reliability coordinator diagnostic support'. This jargon means that the operators did not have real-time data as the system crashed, so they did not know that it was crashing.

How to Prevent Future Blackouts

It's not possible to prevent all future blackouts, but there are valuable lessons to be learnt from the big blackouts of the past.

First, there are advantages to decentralised power generation. One step is obvious — solar cells on the roof.

The second lesson is to realise that redundancy is not always a waste. Having a little slack in the system lets you deal with unexpected emergencies.

Finally, there are disadvantages to relentlessly increasing the speed and connectivity of our essential systems (such as electrical power, Internet access, etc.). A glitch in a super-fast system can make it crash before we have time to recognise the problem.

The Pregnancy Myth

But getting back to the story of blackouts causing a baby boom nine months later ...

This myth has been around for some time. It has been regularly trotted out for many unexpected events where men and women may be caught together, such as ice storms, the September 11 terrorist attacks on the World Trade Center in New York, and the massive blackout of 14 August 2003 in North America. In each case, there was no surge in births nine months later.

While many people futilely awaited statistics to prove a post 9/11 baby boom, Salon.com wisely noted in 2002: 'Ever since the 1930s, Americans have conjured baby booms and busts in the wake of recessions, wars, blizzards and blackouts. Disaster, we reason with amateur zeal, begets increased intimacy, which in turn begets sex' ... and children.

The Blackout Baby Boom

Debunker Man

The debunker of the Baby Blackout Boom theory is Dr S. Philip Morgan, Professor of Sociology at Duke University in the USA and President of the Population Association of America. He's an expert in this field, his research focusing on human fertility.

More specifically, he explores what causes variations in human fertility in different populations. In his two books and 69 published papers in scientific journals, he has studied low human fertility in Iran, the relationship between fertility and the status of women in Nepal, and the link of early childbearing to marriage and subsequent fertility in New Zealand. He has also conducted various other studies in India, Japan, Sudan, Ethiopia, Australia and, of course, the USA.

Professor Morgan says that the Baby Blackout Boom really became an urban myth with the famous New York blackout of

9 November 1965, in which 25 million people were left without power in the USA and parts of Canada. But his studies showed that there was no surge in births nine months later.

Why No Babies?

So what's going on?

There are many factors involved in making babies. Some couples are actually kept separated from each other by the blackout due to commuter problems, a need to stay at work, traffic gridlocks, and the like. Other couples could find the blackout to be a deterrent to sex, e.g. no air conditioning on a hot night. In fact, in the south of the USA, before the advent of air conditioning, the lowest birth rate was nine months after the hottest month of the year. So in hot weather, keeping cool is sometimes more important than getting 'hot'.

Of course, despite the blackout, many couples would still be able to use contraception on the night. (Even with only a candle, people should be able to find a condom.) Some women may be in the infertile part of their menstrual cycle. Even if a couple did conceive, the pregnancy might not go ahead.

And what of the couples who might be forced to stay with nearby relatives and in-laws? That, in itself, can apparently be a powerful anti-aphrodisiac.

Experiment in Air Chemistry

The blackout of 14 August 2003 was massive. Suddenly, 6000 large electrical power plants had to be shut down.

Within 24 hours, the air over the northeastern regions of the USA and parts of Canada became suddenly cleaner. The SO_2 level dropped by 90%, O_3 by 50%, and 'light-scattering particles' dropped by 70%.

The First Blackout

The first recorded blackout happened in New York on 14 October 1889. The headline read, 'A Night of Darkness — More than One Thousand Electric Lights Extinguished'. It was not a technical fault that threw New York into darkness, but an order from an angry mayor.

Over 16 000 light bulbs burnt in New York then — 15 000 in factories and homes, and 1000 on the streets. The electricity was carried to them through high-voltage cables. The contracts had specified that all of New York's electrical supply wires were to be safely buried underground. However, the crooked Tammany appointees to the New York Board of Electrical Control handed out the lucrative contracts to their relatives and friends. They took the money, but just hung the wires above ground, on poles alongside other low-voltage wires already hanging — telegraph, burglar alarms and telephone.

On 11 October 1889, a Western Union linesman accidentally touched one of the live high-voltage wires. According to *The New York Times*, 'the man appeared to be all on fire. There was no movement to the body as it hung in the fatal burning embrace of the wires.'

The Mayor of New York, Hugh Grant, angrily ordered that all electrical power be shut down until the wires were safely buried underground. *The New York Times* reported that 'the sudden disappearance of so many glittering lights through the heart of the city' made 'the aspect of the city ... decidedly provincial'.

References

'A night of darkness: more than one thousand electric lights extinguished', *The New York Times*, 15 October 1889.

'Baby boom debunked', *Australian Doctor*, 28 May 2004, p 20.

Homer-Dixon, Thomas, 'Caught up in our own connections', *The New York Times*, 13 August 2005.

Jonnes, Jill, 'New York unplugged, 1889', *The New York Times*, 13 August 2004.

Low, Marsha, 'Blackout baby boom called an urban myth', *Detroit Free Press*, 13 May 2004.

WHO'S YOUR DADDY?

The word 'paternity' is fairly straightforward. It comes from the Latin word *paternus*, meaning 'relating to a father', and signifies 'the state of being someone's father'. However the subject of 'paternity' is not so straightforward.

It grabbed the headlines in March 2005, when a man who had been put up for adoption 28 years earlier was claimed to be the son of Federal Government Minister Tony Abbott. After much toing and froing, the Australian public was informed that the Minister was not actually the man's biological father.

At the time, the figure of 30% was again bandied about as the percentage of 'fathers' who are not genetically related to their children.

The evidence for this 30% figure is very weak.

30%? Yes — for a Few

In the paternity trade, the technical term used to denote fathers who are not biologically related to their children is 'misattributed paternity'.

It is true that there is an incredibly small subset of Australian men (0.025%) in whom the percentage of misattributed paternity is around 20–30%. (And of course, the media round this figure up to 30%.)

These men have very strong reasons for believing that they are not the biological fathers of their children and get themselves tested by a commercial paternity laboratory. You could call this very small group 'suspicious'.

This figure (around 20–30%) is fairly consistent for various microscopic populations of suspicious fathers all around the world, including the UK, USA, Canada, Australia and Hong Kong.

30%? No — for the Majority

The group of very suspicious Australian men is very small — 0.025%. The remaining 99.975% of Australian men have a very much lower rate of misattributed paternity.

One way to actually measure the percentage of misattributed paternity for the general (and trusting) population is by certain medical conditions that can be passed on to children if, and only if, both parents carry the condition. Typical conditions are cystic fibrosis, retinitis pigmentosa and, oddly, the ability to taste the chemical phenyl-thiocarbamide.

Using this data, the rate of misattributed paternity in the general Australian population is around 1–5%, and usually much closer to 1%. These figures are typical of most Western societies. This 1–5% figure fits in with a 2001 Australian sex survey that looked at 10 173 adults, who had been in a regular relationship for over a year. The survey found that 2.9% of women had more than one sexual partner during the year in question.

One expert in this field is sociologist Professor Michael Gilding, from the Swinburne University of Technology in Melbourne. He has chased down every single bit of public information relating to paternity fraud over the past 30 years, and published his paper 'Rampant misattributed paternity' in the journal *People and Place* in 2005. He wrote that these 30% estimates were 'based on hearsay, anecdote, or published or unevaluable findings'.

Earlier research in this field, such as that done by Drs Macintyre and Sooman, found the same result. In 1991, when commercial paternity testing laboratories were just starting up, they were among the first people to wonder at this often-repeated 30% figure. They wrote that 'high rates have been quoted, but are often unsupported by published evidence or based on unrepresentative populations'.

In the paternity trade, the technical term used to denote fathers who are not biologically related to their children is 'misattributed paternity'. In the 'real world', we call it the 'Milk Man' factor.

They also wrote that 'reliable estimates of the incidence of non-paternity are few and far between'. They were disturbed to find that 'most such references are prefaced by statements such as "it is well known that" or "it is commonly found that" '.

This is Proof?

Where did this figure of 30% for misattributed paternity in the general population come from?

It came from two very unreliable references.

The first was a single throwaway line spoken by Dr Elliot Elias Philipp at a Ciba Foundation Symposium in 1972, on the topic 'Discussion: Moral, Social and Ethical Issues'. This particular seminar was about the ethics of artificial insemination by donor. His comments were about some 'research' done possibly in the 1950s in an unidentified town in southeast England. His exact words were '... we blood-tested some patients in a town in southeast England, and found that 30% of the husbands could not have been the fathers of their children ...'. This single sentence was very controversial (and attractive to the media of the day) and has been regurgitated ever since. However, he did not back up the statement with written research in any peer-reviewed journal, where it could be evaluated by other researchers. The sentence, quoted by the media over and over again, was just a single throwaway line.

The second piece of 'evidence' is the widely quoted 'Liverpool Flats Study' of the 1970s. It supposedly researched non-paternity rates in flats in Liverpool, claiming to find a non-paternity rate of 20–30%. But this 'study' turned out to be nothing more than a few words jotted hurriedly by a student in their lecture notes. No formal statistical study was ever published.

Why 30%?

The 30% figure is very dramatic. The claim that such a large percentage of children are not biologically related to their fathers stuck very effectively in people's memories. Many people wrongly believe that real scientific studies have confirmed that 30% is the non-paternity rate in the general population.

The 30% figure is trotted out endlessly by the media, every time there is any reference to disputed paternity.

It is also publicised by two other groups.

Firstly, there are the support groups for the fathers who are indeed not biologically related to one or more of their children. Finding out that you are not the biological father of your children can be quite distressing, and it's good to have some kind of support in this situation. One support-group web site wrote: 'It is estimated that at least 25% of children living in the Western World aren't the biological offspring of their legal fathers.'

Secondly, there are the paternity laboratories that charge for paternity tests. They have two main clients — unmarried mothers who want to get child support from fathers who deny their paternity, and men who have strong doubts about the fidelity of their partners (or ex-partners). In fact, if you think about it, you would probably expect the non-paternity rate measured in these clients to be much higher than 30%.

Spreading the Myth

The Internet and the media have helped spread this myth. During the media storm around Tony Abbott's 'alleged fatherhood', Professor Gilding's colleague, Dr Lyn Turney, was interviewed in a Brisbane newspaper (*Sunday Mail*, 27 March 2005). She was incorrectly quoted as claiming that the rate of misattributed paternity in the general population was 20%. But she said no such thing to the interviewing journalist. I guess that the old saying 'necessity is the mother of invention' holds true for newspaper editors looking to sell papers.

Look Like Your Father?

How closely do you resemble your parents?

Not a lot, according to Dr Christenfeld from the University of California at Santa Cruz. He reckons, from his study, that by the age of 10, children don't really resemble either of their parents. Younger children don't look like their mothers, but one-year-old children look a bit like their fathers.

References

Aldhous, Peter, 'It's a wise child that looks like its father', *New Scientist*, 16 December 1995.

Gilding, Michael, 'Rampant misattributed paternity: the creation of an urban myth', *People and Place*, 2005, Vol 13, No 2, pp 1-11.

Le Roux, M., et al., 'Non-paternity and genetic counselling', *The Lancet*, 5 September 1992, p 607.

Macintyre, S. and Sooman, A., 'Non-paternity and prenatal genetic screening', *The Lancet*, 5 October 1991, pp 869-871.

CAFFEINE AND BOOZE

Alcohol and coffee are part of human history. Alcohol can be traced back about 7000 years to the village of Hajji Firuz in the Zagros Mountains in Iran. Coffee has a much shorter history, dating to around 850 AD. Tradition has it that Kaldi, a goatherd, was curious about why his goats became so frolicsome after eating beans from the bush *Coffea arabica*. He tried a few of these beans – and that's how coffee entered our cuisine.

So the myth that coffee can sober you up from a state of drunkenness must have appeared some time after 850AD.

Drunkeness and the Bible

Drunkeness is first mentioned in the Bible in *Genesis* 9, with Noah getting drunk, falling asleep without his clothes on and embarrassing his kiddies.

Genesis 9:21 When he drank some of its wine, he became drunk and lay uncovered inside his tent.

Genesis 9:22 Ham, the father of Canaan, saw his father's nakedness and told his two brothers outside.

Genesis 9:23 But Shem and Japheth took a garment and laid it across their shoulders; then they walked in backward and covered their father's nakedness. Their faces were turned the other way so that they would not see their father's nakedness.

And there endeth the lesson ...

Coffee, Alcohol and Driving

Coffee has remarkably few side effects, considering how potent it is. The active ingredient is caffeine, which stimulates your heart, central nervous system, blood vessels and kidneys. Caffeine has a mild diuretic effect, increasing your urine production. It can banish fatigue, keep you alert, improve your motor performance and enhance your senses. It can also make you irritable, jittery and anxious.

Alcohol is also a wondrous chemical. Not only can it preserve dead animals for centuries and remove stains from your garage floor, but in small quantities it is even good for your health. Benjamin Franklin, the great American all-rounder, said that God invented beer to show that He loved us, and because He wanted us to be happy. But while a little alcohol can be good, a lot is bad. It is linked to violence, liver disease and increased rates of car accidents. Countries around the world have laws against driving with too much alcohol in the blood. In Australia, the maximum legal level at which you are allowed to drive a car is 0.05% blood alcohol level.

Unfortunately, this doesn't stop people who have driven to a party sometimes wanting to drive home, even if they have had too much to drink. For this reason, a whole mythology has arisen around how to reduce your Blood Alcohol Concentration (BAC) quickly, so that you can drive home safely (and legally). People try exercise, cold showers, and Kaldi's gift to the human race, coffee.

Coffee, Alcohol and Science

You might think that it makes perfect sense to drink coffee when you're drunk. After all, alcohol makes you slow and sleepy, and caffeine gives you faster reflexes and makes you more awake, so the combination must be good. In fact, many nightclubs in Brazil offer caffeinated energy drinks mixed with alcohol as a cocktail.

Too drunk to ...

Alcohol and coffee are very much a part of human history.

Alch'o'meter™

Alcohol can preserve things immersed in it for centuries,
remove stains from your garage floor
and make a bad band sound GREAT!

Caff'o'meter™

Coffee stimulates your heart, central nervous system,
blood vessels and kidneys ... It can banish fatigue,
keep you alert, improve the performance of your motor skills
and enhance your senses.

This prompted Professor Maria O. Souza-Formigoni from the Department of Psychobiology at the Federal University of São Paulo in Brazil to investigate what happens when you combine alcohol and the beverage Red Bull, an 'energy drink' invented in Austria in 1987. Red Bull contains 80 mg of caffeine, roughly the amount you get in an average cup of coffee. It also has five teaspoons of sugar, as well as glucuronolactone, taurine, pantenol, inositol, various other 'fairy dust' ingredients, and a few vitamins.

In the study, 26 males participated in three different sessions over three weeks. In random order, they drank alcohol by itself, then Red Bull by itself, and then a mixture of alcohol and Red Bull.

The Results

The research results were startling.

First, the caffeinated energy drink, Red Bull, did not speed up the metabolism of alcohol. On average, your Blood Alcohol Concentration drops by 0.015% per hour, and there's not much you can do to speed it up.

Second, the volunteers did claim that they felt more awake and alert, and less headachy. They also claimed that they had fewer instances of dry mouth, and less weakness and impairment of motor coordination. However, the objective measurements showed the opposite. In reality, the caffeinated energy drink did not improve their motor hand–eye coordination or their reaction times.

In other words, the drunken, caffeine-loaded volunteers were deluding themselves — because the alcohol not only influenced their motor skills but also their ability to make rational decisions.

This highlights the real problem with taking caffeine to sober yourself up. A drunken, un-caffeinated person knows that they are drunk, and hopefully won't drive. A drunken, caffeine-charged person is more likely to believe that they are fine — and more likely to grab their car keys and try to drive.

So, if deciding whether to drive or not, caffeine can be the red rag to the bull.

Other Sobering-up Myths

Some people recommend a cold shower as a way to sober up quickly. But all it gives you is a cold and wet, drunken person.

Other people recommend vigorous exercise — which gives you a tired, drunken person.

Artificial Sweeteners Make you Drunker

Dr Chris Rayner from the Royal Adelaide Hospital compared the effects of alcohol mixer drinks, when the mixer was based on either sugar or an artificial sweetener.

The alcohol mixed with artificial sweetener made the participants drunker — 0.05% Blood Alcohol Concentration, compared to 0.03% for the sugar-based drinks. The peak in BAC happened at the same time for the artificial sweetener and the sugar, but it was a lot bigger for the artificial sweetener.

Why? Artificial sweeteners accelerate the emptying of the stomach. Once the alcohol has left the stomach, it can be better absorbed by the next section of the gut.

So when you are next at a cocktail party, factor in the effect of the sweetener.

Caffeine Makes You Say 'Yes'

It's well known that alcohol can remove your inhibitions to perform certain activities. Dr Pearl Martin of the School of Psychology at the University of Queensland says that caffeine can also make you say 'yes' more often. I wonder what happens in drinks that combine both alcohol and caffeine?

Perhaps we need to update Ogden Nash's poem:

Candy is dandy,
Liquor is quicker,
But coffee gets you hotter to trotter ...

References

Encyclopaedia Britannica, Ultimate Reference Suite DVD, 2006 — 'caffeine'.

Souza-Formigoni, Maria O., et al., 'Effects on energy drink ingestion on alcohol intoxication', *Alcoholism: Clinical and Experimental Research*, April 2006, pp 1-8.

SPOT REDUCTION OF FAT

Haven't most of us dreamt of being able to 'spot reduce' fat? With a few specific exercises, we hope to remove fat from the waist (usually men) or hips and buttocks (usually women). It's a beautiful dream — but no more than that.

Fat 101

When you eat fats they are broken down by a chemical called lipase into smaller chemicals, such as glycerol, fatty acids and glycerides. These smaller chemicals, then pass through the wall of the gut into the blood, where they are recombined. Microscopic droplets of fat then travel through the body to be used or stored.

When fat is stored it acts as an insulator, a shock absorber, and, of course, a fuel reserve. Fat is a terrific storehouse of energy. On a weight-for-weight basis, it stores twice as much energy as either carbohydrate or protein.

Fat Storage

The body's fat storage locations are predetermined by your genes and your gender.

Behold ...
'The handles of love' ...

When stored, fat acts as an insulator, a shock absorber and, of course, a fuel reserve.

For illustrative purposes, we have used a dude.

Where fat is stored though, is predetermined by your genes and your gender.

In general, men tend to build up fat around the waist — creating the 'apple' shape. This external abdominal fat mirrors fat around the organs inside the belly. Medical statistics show that this pattern puts people at higher risk of diseases such as heart disease and diabetes.

On the other hand, women tend to build up fat around the hips and buttocks — creating the 'pear' shape. This is called 'reproductive fat' — the emergency supply of energy needed to grow a baby. It doesn't have the same health risks as abdominal fat.

Spot Reduction

Many advertisements claim to help you 'spot reduce' your fat away by offering exercises and 'biomechanically correct' devices. Most impressively, the advertisements are always presented by slim models with striking abdominal and buttock muscles.

If you perform several hundred sit-ups per day, you still don't burn the fat on your tummy. You just get stronger abdominal muscles. If you don't change your eating or cardiovascular exercise habits, those stronger abs will just be your little secret buried under your tummy fat. In general, when lots of fat is stored in one area, it was usually the first to be laid down, and will be the last to fade away when you eat less and exercise more. Your lovely 'washboard abs' will be visible only when you get your total body fat down to very low levels.

Shawn Phillips, the fitness trainer who has advised Sly Stallone and Brad Pitt, says in his bestselling book, *ABSolution: The Practical Solution for Building Your Best Abs,* that the myth about the spot reduction of fat is the cause of many tragic cases of fitness failures. (This falls into the category of a non-Shakespearean tragedy.) And Arnold Schwarzenegger, seven-times Mr Olympia winner, agrees. He writes in his *The New Encyclopedia of Modern Bodybuilding*: 'When the body begins metabolising fat for energy, it doesn't go to an area where the muscles are doing a lot of work to get additional energy resources.'

Not to take anything away from having good tummy muscles ... They will give you better posture, and probably make you happier to pull on your swimming costume at the beach.

Fat is a superb source of fuel. In order to burn enough kilojoules to lose a kilogram of fat you need to do about a million sit-ups. Even Elle 'The Body' Macpherson doesn't come close to that ...

Tummy Insecurity

'Tummy insecurity' is not new. Both men and women have pointed to the tummy for decades as the one area of the body that they would like to improve. In the old days, the shamsters would advocate vibrating belts, sauna sweat shorts, belly creams, cling wrap plastic, etc. to tone the tummy. Of course, none of them worked.

As the consumers got a little smarter, the shamsters got smarter by marketing new rubbish. In the mid-1990s, the Ab Roller Plus was marketed with the slogan 'Get a flat, sexy stomach in five minutes flat!'. One American home-shopping channel, *QVC*, reported that in one lovely 15-hour window, they sold 41 000 Weider Ab Shapers (about one per second) at US$40 each.

References

Kuntz, Tom, 'Cashing in on the abs obsession', Ideas and Trends, *The New York Times*, 9 June 1996.

Schwarzenegger, Arnold with Dobbins, Bill, *The New Encyclopaedia of Modern Bodybuilding*, New York: Simon & Shuster, 1998, p 538.

RIP-ROARING TIME

A few years ago, three generations of my extended family (eight of us) rented a house for a fortnight at Forster, a few hours drive north of Sydney and close to the beach. It was in the ocean, at the southern end of the beach and only a little way outside the flags, that I had yet another brush with death. I was swimming with my 13-year-old son, who is a stronger swimmer than I am. We were out past the breakers in the smooth water, when I suddenly realised that we were caught in a rip. The shore kept getting further away, no matter how hard I tried to swim back towards it. I began to get very scared.

Like most people, I wrongly believed that rips are rare and unusual events, and that once you get caught in one, you're on your way to a far distant shore.

Rips are Normal

Rips are absolutely normal on every beach. You find them on every beach. Dr Rob Brander (previously at the University of New South Wales) says that rips are a major mechanism for the seaward transport of water and sediments. They have a strong effect on the shape of the sea floor near the beach and help disperse pollutants out to sea. Unfortunately, rips are also a major hazard to swimmers.

A rip (called an 'undertow' in the UK) is a narrow, concentrated channel of water flowing out to sea, i.e. like a river in the ocean.

They are part of the normal water circulation pattern of every beach. Rips are how the water that gets carried in by the waves gets back out to sea. Once the water has come in, it has to go out. (If the incoming water didn't go out, it would pile up very rapidly on the shore.)

If you have a fixed structure like a cliff, reef or pier, a permanent rip can establish itself right next to it. However, on sandy beaches, rips tend to move around over time.

Rips flow out to sea through channels in the sand. These channels can expand and contract with the surf conditions, the tide and the weather. It usually takes a really large storm to physically shift the channel in the sand to a different location. Rips have been known to continue at the same location on a beach for weeks and even months.

They are usually weak, but occasionally they will become strong — and dangerous.

And on a beach that is long enough, there can be many rips, each roughly the same distance apart.

Coastal scientists have been studying rips for three-quarters of a century. There are many theories as to why a rip channel should form in any particular location — but none of the theories provide a full explanation.

How a Rip Works

The mechanism of the rip is easy to understand.

As the waves head for the shore and break into white water, they bring the water in towards the shore. The water runs up the sloping sand and then rolls back into the shallow water a little way, where it is stopped by the incoming waves. This forces the returning water to turn (say) to the north and flow parallel to the shoreline. It is now called a feeder current. After a hundred metres or so, it will run head-on into another feeder current running (say) south. The two feeder currents then merge and head out to sea. They have merged to form what scientists call a 'rip current' or rip.

As it heads seawards, the rip quickly consolidates to make a narrow, fast-moving neck, usually about 10 m wide (although it can vary between 3–20 m across). All the water in the rip (from the surface down to the sandy bottom) is moving seawards, although it is slowest down near the sand. The rip keeps its structural integrity and shape, until it gets past the breakers and out into the unbroken rolling waves. The neck then widens into a broad 'rip head', the moving water gradually slowing down until only the water at the surface is moving seawards. When it too finally stops, so does the seaward motion of the rip.

The water in the rip head spreads sideways and is carried in towards the shore by the next wave, beginning the cycle again. In some cases, the rip will actually carry you back to shore. So the fear that you'll get carried way out to sea or to a distant shore is totally unfounded — but in a panic situation, you'll believe anything. When I was caught in the rip, I was very scared.

Find a Rip

Before you plunge into the surf it's always a good idea to spend a few minutes checking the area for rips. There are a number of different ways to spot a rip — and you get better at it the more you look.

The easiest way to find a rip on a patrolled beach is to check out the sand, not the water. Look around for the big fat rescue boards — the lifesavers usually set them on the sand directly opposite a rip. But this method only works at patrolled beaches.

You can also look for persistent bands of dark water heading offshore. Rips tend to go out in deeper channels in the sandy bottom — the water is darker, because it's deeper. Fewer waves break there, so there is less white water. This situation is particularly tricky. The water might look safer because there are fewer waves breaking there but it's actually more dangerous. The surface of a rip is often choppy and disturbed, because the rip is heading out to sea against the direction of the incoming waves.

Finally, you can sometimes spot a rip because it is carrying stuff out to sea — sand, seaweed, jellyfish, people, pets, etc.

Many overseas visitors get caught in rips because they don't know what a rip is, let alone what to look for. Some of them believe,

'Ripper!'

Rips are an absolutely normal function of all beaches and are part of the normal circulation pattern of every beach. Simply put, the waves carry the water in, and the rips carry it back out to sea.

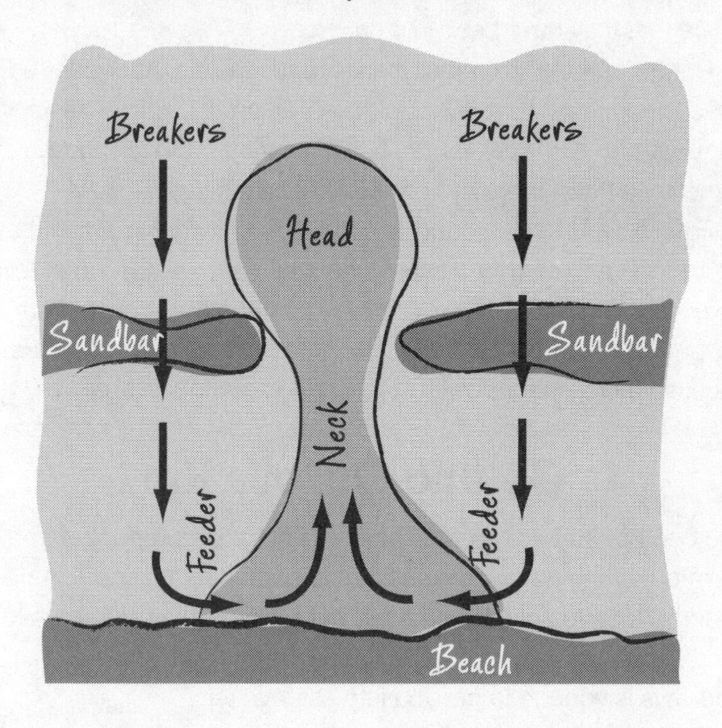

As the waves head for the shore and break, they bring in water. The water then needs to run back out to sea, but is stopped by more incoming waves. The water now runs parallel to the shore trying to find a path of least resistance so as to flow back out to sea ... this is called a feeder current. When two feeder currents merge (or one finds a fixed structure) the water then heads out to sea. This is called a rip current.

from their European experience, that the flags designate a private beach area, and so they head for the unpatrolled waters — and straight into a rip. Some days during summer, lifesavers at Bondi Beach rescue more than 100 people from rips.

Speed of Rips

Sometimes, the water in the rip flows so slowly that it's very easy to swim against or across it. The average swimmer might not even notice that they have been in a rip.

But at the other extreme, there are mega-rips. At Sydney's Palm Beach mega-rips have been clocked at over 7 kph — faster than an Olympic swimmer in a 50 m sprint. (These unusual rips sometimes travel more than one kilometre offshore.)

Rips flow fastest around low tide. The outgoing water has to return to sea through a smaller 'channel' because the water level is lower. (In southeast Florida, three-quarters of rescues from rips happen at low tide.) Rips are also faster when the waves are big, because there is more water to shift back out to sea again.

Getting Out of a Rip

You can't swim back to the shore against a strong rip, but you can swim out of the rip by going across it, parallel to the shore. A rip is a treadmill that you can't switch off, but by swimming sideways to the shore you can get out of it. Sometimes, swimming just 10 m sideways is enough to get you into safer water.

Back to me, struggling in a rip. My son advised me to stop swimming against the rip current and exhausting myself. He told me to tread water instead using the 'eggbeater kick', which he was doing effortlessly. I had never heard of this, but still managed to calm down and float as the rip carried us further out. I didn't know it, but back on the shore my wife and the lifesaver were already discussing whether to mount a rescue for me and my son. However, I hadn't started waving my arms to signify that I wanted to be saved (stupid male pride!). Suddenly, we realised that we were being carried towards some surfboard riders.

Saved! We swam across to a 14-year-old kid and I asked if I could hang onto his board and get my breath back (my son didn't need to). After a few minutes, we swam a little way out of the rip and caught the waves back to shore.

So when you're next at the beach, pay attention, and you'll have a rip-roaring time …

Gone in 60 Seconds — or 20 …

On average, a drowning adult can wave their arms for help for about 60 seconds in still water and for much less time in choppy water — a very small window of time to be spotted by a lifesaver.

But a drowning child can wave their arms for only one-third as long, which is why you need to keep an eye on kids in the surf.

American Rip Stats

In the USA, according to the United States Lifesaving Association, lifeguards rescue over 100 000 people from drowning each year.

They estimate that 80% of surf rescues (over 22 000 each year) happen because of rip currents. More than 100 people die each year because of rips.

One analysis found that there were more surf rip rescues when the waves were coming in square to the beach, when the tide was low, when the wave height at sea was between 0.5 and 1.0 m, and when the waves were 8–10 seconds apart. This is probably more maths than the average swimmer is prepared to do before diving into the surf.

Rip Pulse

If there is a set of unusually large waves, you can experience a 'rip pulse'. Here the rip speeds up to funnel the water back out to sea again. A rip can double its speed in seconds — with disastrous results.

A rip pulse happened at Bondi Beach in 1938. An unexpected set of three huge waves swept 300 swimmers across into a rip. Because of the extra water it had to shift from the big waves, the rip was a lot faster for a few moments — long enough to take the 300 swimmers out to sea. Eighty members of the Bondi Surf Bathers' Lifesaving Club swung into action. Thanks to their heroic efforts, most of the swimmers survived, and 40 were brought back to shore unconscious. Sadly, five died.

So be warned — rip pulses are very dangerous. A normal rip on a sandy beach may carry you 100 m offshore, but a rip pulse can take you out several hundred metres before it runs out of grunt.

World Drowning Stats

According to the World Health Organisation, in 2000, approximately 410 000 people drowned. This makes drowning the second leading cause of unintentional death after road traffic injuries.

These drownings happened mostly (97%) in low- and middle-income countries. In 2000, drowning killed 3458 people in the USA and 322 people in Australia.

Non-oceanic Rips

Rips can even occur in lakes, if they're big enough. In the USA, in the years 2002 and 2003, 18 people drowned from rips in Lake Michigan alone.

Sometimes it's Hard ...

A Bondi surf lifesaver told me about the difficulty he had rescuing a really fat man who was caught in a rip.

'The fat guy was absolutely covered in sunscreen. He was so slippery that every time my mate managed to shove him onto the rescue board, he slipped off again. My mate had to call for another two rescue boards and eventually straddled the fat guy across the three rescue boards side by side, with the lifesavers holding him on.

'Once they got him back to shore, he talked to a friend for a few minutes and then plunged into the surf again. At least this time he kept between the flags.'

References

Branche C.M. and Stewart S. (Eds), 'Lifeguard Effectiveness: A Report of the Working Group', Atlanta: Centers for Disease Control and Prevention, National Center for Injury Prevention and Control, 2001.

Engle, Jason, et al., 'Formulation of a Rip Current Predictive Index Using Rescue Data', Proceedings of National Conference on Beach Preservation Technology, FSBPA, 23-25 January 2002, Biloxi, Mississippi.

Haller, Merrick C., et al., 'Experimental study of nearshore dynamics on a barred beach with rip channels', *Journal of Geophysical Research*, 2002, Vol 107, No C6, 10.1029/2001JC000955.

Haller, Merrick C. and Dalrymple R. A., 'Rip current instabilities', *Journal of Fluid Mechanics*, 2001, Vol 433, pp 161-192.

http://www.ripcurrents.noaa.gov

GIVE PEACE A CHANCE

On New Years Eve, our whole family caught the bus into town to see the fireworks. The mood on the bus was happy, bubbly and convivial. Across the aisle, I could see my daughter and her girlfriend chatting but I couldn't hear what they were saying (they were too far away and there was a low level of background noise). In another seat nearby I could see my parents-in-law having an animated discussion — but again, I couldn't hear what they were saying. Then, right at the back of the bus, a mobile phone rang and a man answered it. And everybody on the bus could hear everything he said, beginning with the predictable shouted line, 'I'm on the bus'.

He didn't realise it, but he could have actually spoken at a normal conversation level, and not annoyed everyone else. Two technological factors were involved in why he shouted unnecessarily — the presence of Automatic Gain Control and the absence of Side Tone.

Automatic Gain Control

Let's start with Automatic Gain Control, or AGC.

The first AGC circuits were invented in the 1920s and 1930s, in the early days of electronics when glass valves ruled the roost, 50 years before the introduction of integrated circuits and transistors.

'Can you hear me?'

Most people don't realise that you can speak at normal conversation levels on a mobile/cell and still be heard perfectly ... so ... STOP SHOUTING!!!

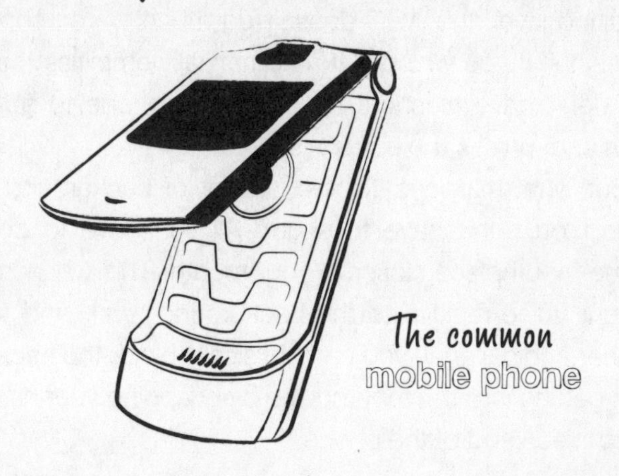

The common mobile phone

Nearly all mobiles have Automatic Gain Control (AGC). It's fancy jargon for 'amplification' and means that when you yell, it reduces your voice and when you whisper, it amplifies it.

'Gain', as in Automatic *Gain* Control, is fancy electronic jargon for 'amplification'. We need automatic amplification in our various electronic receivers (TV sets, mobile phones, radios, etc.) because we all get different strengths of signals in our antennae. For example, your favourite TV channel might pump out 100 000 watts of power from its TV broadcasting antenna.

This power is broadcast in all directions, so very little of it gets to your antenna. If you live fairly close to the broadcast antenna, your home TV antenna might pick up one-hundredth of one-millionth of a watt (0.000 000 01 watts). This faint signal will be amplified by the electronics inside your TV set. However, if you live 80 km from the broadcast antenna, your home antenna might receive a signal 1000 times weaker (only 0.000 000 000 01 watts). This weaker signal has to be amplified 1000 times more than the signal of the person who lives close to the broadcast antenna. But you, the home viewer, do not have to adjust the internal video amplifier in the TV set. The amplifier will do this for you automatically via the Automatic Gain Control.

In exactly the same way, an AGC circuit controls the signal from the mouthpiece in your mobile phone. If you talk at a normal speaking level, the AGC doesn't do much. If you shout, it reduces the signal. If you whisper, it automatically amplifies the signal more. You don't have to shout into a mobile phone to be heard at the other end (unless the battery is dead).

But what happens if there is a lot of background noise? Once again, you don't have to shout. All you have to do is bring the phone mouthpiece closer to your mouth. This will increase the level of your voice relative to the background level. And to send out a cleaner audio signal, you can physically block the background noise from getting into the mouthpiece by wrapping your hand around the mouthpiece end of the phone.

Side Tone

The second technological factor is called Side Tone. The Institute of Electrical and Electronics Engineers has a *Standard Dictionary of Electrical and Electronics Terms*. It defines 'Side Tone' as 'the acoustic signal resulting from a portion of the transmitted signal being coupled to the receiver of the same handset'. In other words, when you speak into the mouthpiece a small part of the signal is sent to your earpiece. If you can hear your own voice in the earpiece, you can be sure that your signal is getting out. You can now adjust your speaking level to match your listening level.

Now here's the weird bit. Those old-fashioned landline phones have Side Tone. The new 'fancy schmancy' mobile phones do not!

So people on mobiles don't hear any of what they're saying coming through the earpiece, and they assume that their speaking level is too low. And that's the second reason why they shout.

Flying Mobiles

At the moment, these annoying interludes — having to listen to somebody shout out the trivialities of their lives and times to all and sundry — are relatively short.

However, the airlines are just finishing technical studies on allowing people to use mobile phones inflight (the airline business has small profit margins and they want the extra money). In planes, the seats are narrow, the food is sometimes dubious, but at least you get some silence. On a long flight do you want to listen to someone complaining on the phone about their ex, or to a salesman ringing every single contact to sell them aluminium siding?

Perhaps flight attendants will have to include AGC and Side Tone Awareness in the safety demonstration before every flight ...

Ear and Earpiece

There's a third, and rather silly, factor that can make you shout on the phone. I know because it happened to me several years ago. If you can't hear the other person clearly, you are also likely to shout on the phone.

A little while after I had bought a new mobile phone I began to notice that I was speaking rather loudly, but didn't really pay it much attention. If I had thought about it, I would have realised that it was because the signal coming out of the earpiece was so quiet.

One day I happened to hold the phone in a slightly different way — and suddenly the voice level in my ear was so much louder. For several months I had been holding the earpiece speaker a little off to one side instead of directly against my ear!

Getting Rid of Background Noise

One problem that audio engineers have to deal with is 'background noise'. Fancy electronic tricks can get rid of this problem.

The Australian Navy solved the problem by placing two separate microphones a short distance apart. Suppose there was a battle going on, with all the noise that you would expect — shouts, small arms fire, explosions, guns, etc. Both mikes would pick up the background noise. But the person would speak into just one microphone. One mike would only have the sound of battle, the other mike the sound of battle plus the voice. The two mikes would then be fed into an electronic circuit that would subtract one mike from the other, leaving behind just clean voice. Fairly primitive, but surprisingly effective.

AGC Everywhere

Automatic Gain Control circuits are in virtually all consumer electronic products and most professional ones.

They are in VCRs and DVD recorders to give a signal that is large enough to record well, but not so large as to 'saturate' the recording medium and cause distortion. When a DJ switches between various sound inputs (MP3 file, CD player, record player, etc.), the AGC compensates for the different signal levels from each of these inputs. In cable TV, the signal gets weaker as the cable gets longer, so at certain fixed distances there are AGC boxes to pump up the signal.

Yes, AGC is everywhere, including your mobile phone.

Locust Hearing

If a locust cannot hear the smallest of noises, it might get eaten, or miss out on dinner. Locust ears have membranes, millionths of a metre thick. When they listen to very quiet sounds, the membranes vibrate back and forth billionths of a metre (a human hair is about 80 000 billionths of a metre across).

Professor Daniel Robert from the University of Bristol has discovered that locust ears (like human ears) use AGC to vibrate the membranes automatically, and so make the incoming sound effectively louder. We still don't fully understand how locust (and human) ears work, but if we ever do we can use this information to make new microphones able to pick up extremely soft sounds.

References

Encyclopædia Britannica, Ultimate Reference Suite DVD, 2006 — 'radio'.

Encyclopædia Britannica, Ultimate Reference Suite DVD, 2006 — 'television'.

Stein, Ben, 'Cellphones in flight? This means war!', *The New York Times*, 26 March 2006.

USE YOUR NOODLE

Noodles are a big hit with most kids, and with the parents who are trying to feed them. Most of the names of the various noodles are Italian. *Ravioli* means 'little envelopes', *cannelloni* 'big tubes', *linguine* 'little tongues', *tortellini* 'little cakes', *lasagna* 'baking pot', *vermicelli* 'little worms', while the word *pasta* means 'dough' or 'paste'. With all these Italian names, it's easy to believe that noodles originated in Italy — but they didn't.

Early Agriculture

Feeding ourselves has always been very important. At some stage, we made the transition from foraging in the wild to growing crops under cultivation — we're not sure exactly when this took place.

We do know that 27 000 years ago, stone tools were used in the Solomon Islands to work taro. Some 20 000 years ago, at the Paleolithic Ohalo II site in Israel, our ancestors were using stone tools to pound, grind and bake the large hard fibrous seeds of wheat and barley to make them easy to digest. It was more economical in terms of labour and food to use the largest seeds they could find.

The First Noodles

The first known appearance of noodles was about 4000 years ago in northwestern China, on a terrace on the upper reaches of the Yellow River. Archaeologists call it the Lajia Neolithic Settlement. The geological evidence seems to show that a huge earthquake, followed by catastrophic flooding, destroyed the settlement.

In the debris, archaeologists found a prehistoric bowl of noodles, upside down under 3 m of brownish-yellow, fine clay floodplain sediment. The noodles were sitting in the small air gap on top of the clay that almost filled the upside-down bowl.

Dr Houyuan Lu, from the Institute of Geology and Geophysics at the Chinese Academy of Sciences in Beijing, wrote in *Nature*: 'The

Using your noodle ...

The first noodles appeared about 4000 years ago in northwestern China.

There is a major difference between noodles and pasta. Noodles are mostly made from bread wheat and usually eaten in something like soup. Pasta is mostly made from durum flour and usually eaten in dishes containing limited water.

noodles were thin (about 0.3 cm in diameter), delicate, more than 50 cm in length and yellow in colour. They resemble the La-Mian noodle, a traditional noodle that is made by repeatedly pulling and stretching the dough by hand.'

Noodles of Millet

His team used two different methods to identify the plants from which the noodles were made. They found that they were made from the cereal, millet.

First, they looked at the starch granules through a microscope and found that the granules were the size (about 10 microns, roughly one-tenth the thickness of human hair) and shape of starch granules found in modern millet.

Second, they found that tiny phytoliths in the noodles also pointed the finger at millet. 'Phytoliths', unique to each plant, are tiny pieces of glass that some plants make. The glass is not optical-grade transparent glass, but it *is* glass. Perhaps it's not so surprising. Both rocks and glass are made from silica and both rocks and plants exist together in the soil. Some plants have worked out how to extract the silica from the soil to make glass. And, strangely, they can do this without resorting to the high temperatures that human beings have to use. Indeed, the stinging tips of the nettle plant are tiny pointy sticks of glass.

Travelling Noodles

How did these noodles get to the rest of the world?

The Japanese Envoy to China introduced Chinese noodles to Japan during the Eastern Han Dynasty (25–200 AD). Tradition has it that Marco Polo (and his dad, Niccolo and uncle, Maffeo) returned from China at the end of the 13th century with noodle recipes and preparation tools for the chefs of Venice. Since then, Italians have developed many different types of pasta.

In 1353, Boccaccio's *Decameron* (*Il Decamerone*) was published. The book is a collection of 100 imaginative stories that a group of quarantined citizens of Florence told each other over a period of

10 days. Hence the name of the book — *decamerone* means '10 days'. And pasta is mentioned in the book: 'In the region called Bengodi, where they tie the vines with sausage, there is a mountain made of grated parmesan cheese on which men work all day making spaghetti and ravioli, eating them in capon's broth.'

There's a major difference between noodles and pasta. Noodles are made mostly from bread wheat and are usually eaten in a water-based meal, such as soup. Pasta is made mostly from durum wheat, and is usually eaten in dishes containing a limited amount of water.

So if you thought that the Italians were faster with the pasta, you'd be off your noodle — because the Chinese got there first.

Fasta Pasta

Why do we add salt (NaCl) to the water in which we cook pasta? There are two answers, but one of them is wrong.

The correct answer is that the salt enhances the flavour of the pasta.

The wrong answer is that adding the salt increases the boiling point of the water — the higher temperature supposedly cooks the pasta faster. There is a small element of truth here. Adding 20 g of salt (a tablespoon) to 5 l of water will increase the boiling point — from 100°C to 100.04°C. The increase is pretty microscopic and will hardly cook the pasta any faster.

Durum Wheat is Not Regular Wheat

The scientific name for regular or common or 'bread' wheat is *Triticum aestivum*. Its DNA is arranged as six sets of chromosomes (hexaploid).

Durum wheat is a different species, with the scientific name *Triticum turgidum* (actually its real name is *Triticum turgidum var. durum*). Its DNA is arranged into four sets of chromosomes (tetraploid).

Noodles vs Pasta

The vast majority of noodles are Asian wheat-based noodles. China makes over 100 million tonnes of noodles each year — and still has to import more to satisfy the demand. These noodles are made by mixing bread-wheat flour with salt and water. The 'salt' can range from 'regular' salt (NaCl) to 'alkaline' salt (Na_2CO_3 or K_2CO_3). This mix gives a dry crumble that is then formed into a continuous sheet, which is then cut into strands to make the final noodles.

Pasta is usually made with durum wheat (which is not really a wheat). It is then mixed with water and drawn or moulded to make the desired shape.

References

Lu, H., et al., 'Millet noodles in late Neolithic China', *Nature*, 13 October 2005, pp 967-968.

Moore, Peter D., 'Getting to the root of tubers', *Nature*, 24 September 1999, pp 330-331.

Panati, Charles, *The Extraordinary Origins of Everyday Things*, New York: Harper & Row, 1987, pp 405-406.

Piperno, D. H., et al., 'Processing of wild cereal grains in the Upper Palaeolithic revealed by starch grain analysis', *Nature*, 5 August 2004, pp 670-673.

Wolke, Robert L., *What Einstein Told his Cook*, New York: W. W. Norton & Company, 2002, pp 46-48.

MICROWAVE OVEN NOT RADIOACTIVE

I was heating my lunch (I love leftovers) in the microwave oven in the Physics tearoom when a visiting Chinese professor came in to heat up her meal. I introduced myself and told her that my food would be out of the nuker in less than a minute, so she could shortly, in turn, nuke hers. We then had a confusing conversation. The confusion ended when I suddenly realised that the word 'nuker' was not part of her vocabulary. I explained that in Australia, you can 'nuke' your food in the 'nuker' to heat it up, and that the origin of the word was the nuclear reactor. Rather puzzled, she replied, 'But there is nothing nuclear about microwaves.'

She was right. When I give a public talk, I sometimes ask the audience to name something radioactive that is found in most homes. About half of them will name the microwave oven. But they are wrong. The most common radioactive device in our homes is the smoke detector.

There is absolutely nothing radioactive about a microwave oven. It does not generate radioactivity, nor does it have any radioactive parts — unless some wild inventor incorporates a smoke detector into one!

Radioactivity Eureka

The word 'radioactivity' was coined by the Nobel Prize winner Marie Curie in 1899.

Radioactivity had been accidentally discovered by her fellow Nobel Prize winner, Antoine-Henri Becquerel, a few years earlier, in 1896. Becqueral was following up on the recently discovered type of radiation called x-rays. He was studying this new field of invisible radiation to see what happened to photographic film that was protected from light by an opaque wrapping. (For example, whether x-rays would penetrate the wrapping and leave their mark on the photographic film, even though it had never been exposed to light.) By chance, he left some uranium salts on top of some wrapped photographic film in a closed drawer. When he processed the film, he could clearly see the imprint of the uranium salts. The uranium had emitted a mysterious radiation that could penetrate paper and expose film. Becqueral had discovered radioactivity.

Most radioactivity happens when a big atom splits into smaller atoms or when small atoms combine to make a bigger atom. As part of this process, subatomic particles are given off. Energy is also given off, usually as ElectroMagnetic Radiation (EMR). We call this combined emission of subatomic particles and energy 'radioactivity'.

Ionising Radiation

For now, let's ignore the subatomic particles and look only at the EMR. The higher the frequency, the higher the energy.

In many cases of radioactivity, the frequency of the electromagnetic waves emitted is high enough that they can damage atoms by knocking electrons off them. The atom is now called an 'ion' — and the radiation is called 'ionising radiation'. This kind of high-energy EMR, such as gamma rays or x-rays, can easily damage flesh and bone, organs, and DNA.

Microwaves, however, have a relatively low frequency, and so are low-energy EMR. Microwaves do not have enough energy to ionise atoms. They are indeed electromagnetic radiation, but are not radioactive, even though they are a type of radiation. After all, the beam of light that comes out of your flashlight is a type of radiation, but light is not radioactive.

Non-ionising Radiation

The microwave oven does, in fact, generate electromagnetic radiation. These waves of EMR happen to be in the microwave frequency band, which is why we call them microwaves. Each wave is about 10 cm from one peak to the next.

The microwaves give the water molecules in food a real shake-up. Each water molecule looks like a little letter 'V', with the oxygen atom at the point of the 'V', and each of the two hydrogen atoms at the ends of the arms of the 'V'. As each wave of the microwave radiation rushes past, the water molecules swing back and forth around the point of the 'V'. At the same time, the arms get shorter and longer as they shrink and expand. Down on the molecular level, more movement means more heat. All this molecular movement creates friction, which generates heat. The heat moves directly from the water to the fats and proteins. This is how a microwave oven heats food without melting the plastic container, which has no water molecules in it.

Now here's the essential point. Yes, x-ray EMR has enough energy to knock electrons off atoms — which is one way that it damages flesh. But the microwave EMR does not have enough energy to knock electrons off atoms.

Other Non-ionising Radiation

If we go to the next, more energetic band of EMR (infrared or, in plain English, heat) we find the same situation. The infrared EMR that a campfire or a stove emits has more energy than microwave EMR. However, it doesn't have enough energy to knock electrons off atoms. Even so, heat can raise a nasty blister on your skin.

Going up the EMR frequency and energy scale, the next form of radiation is visible light. If the light is bright enough, it can blind your eyes, but once again it doesn't have enough energy to knock electrons off atoms.

The Electromagnetic Spectrum

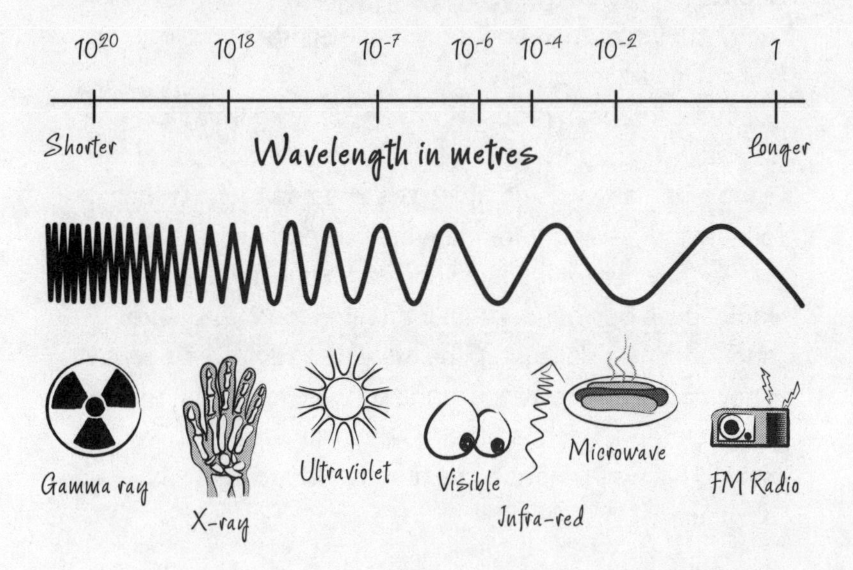

But the next EMR does. Called ultraviolet light, it is known to cause skin cancers by interfering with the DNA repair mechanism.

Gutless Microwaves

Microwave radiation has less energy than heat, visible light and ultraviolet light. Few of us are fearful enough to carry umbrellas to shield us against the dangerous ultraviolet radiation from the Sun. But many of us have a vague, poorly defined fear of microwaves (which have less energy than ultraviolet EMR) and microwave ovens. Perhaps some people are scared of change — I can see why.

Using microwaves is the first really new way of cooking food to come up in the past million years, apart from frying in oil and heat-resistant pots, which probably appeared 5–10 000 years ago.

Thanks to the microwave oven I don't have to see all those lovely leftovers sitting so cold and miserable, and going to waste ...

Strange First Use of Radioactivity

I remember reading this story in the *New Scientist* in the early 1980s (or thereabouts). Here it is, from memory (which is very easy to fool).

In the 1920s, a group of physicists were living together in a boarding house in England. They suspected that the stingy landlady was scraping the uneaten food off their plates and reusing it to make their weekly stew.

So they surreptitiously sprinkled some low-level radioactive powder over their uneaten food. And sure enough, a few days later, the stew was slightly radioactive. They had proved that the landlady was 'recycling' their uneaten food, because the radioactivity turned up in their stew. This was probably the first use of radioactivity — as a tracer.

Atoms 101

The word 'atom' comes from the Greek, via Latin and Old French. It means 'something that cannot be cut or divided into anything smaller and still be itself'. For example, a kilogram of butter is butter, as is a gram of butter. Butter is fat with water added to it. But if you keep on cutting the sample of butter smaller and smaller, at some stage it won't be butter any more. You'll get down to a molecule of fat, and a molecule of water.

The word 'atom' has two parts — 'a' meaning 'not' and 'tom' from *temnein*, meaning 'to cut'.

So the name is slightly wrong, because you *can* cut an atom into smaller parts. At this stage, you now have subatomic particles (e.g. electrons and protons).

Up to the present time about 110 different atoms have been identified, beginning with hydrogen (the lightest) and going up past uranium (at 92) to darmstadtium (at 110). Their mass ranges from 1 (hydrogen) to 281 (darmstadtium).

Atoms are mostly empty space. They have clouds of electrons (hardly any mass) around a central nucleus (where 99.9+% of the mass of the atom is). The nucleus is very small compared to the outer electron cloud — roughly equivalent to a marble in a football field. The nucleus takes up about one hundredth of one-trillionth of the total volume of the atom.

Surprisingly, even though their masses vary enormously, all atoms are roughly the same size — about 0.2–0.4 nm (a nanometre is one-billionth of a metre).

Again surprisingly, the diameters of all the nuclei are quite similar — 4–15 fm (a femtometre is one-millionth of one-billionth of a metre).

Radioactivity Units 101

The units used to measure radioactivity are a little confusing. Some of them just measure how many atoms get 'split', while others try to measure how much damage the radiation does to a human being — not any other living creature, just a human being. So I'll use a metaphor from the martial arts.

First, there is the Becquerel (Bq). Officially, this unit measures how many atoms split per second, or how many nuclei disintegrate each second. In my martial arts metaphor, the Becquerel is a measure of how many blows get thrown — regardless of whether they are kicks or punches, or what type of kick or punch, or even if they hit you or completely miss you.

Second, there is the Gray (Gy). In technical talk, it measures how much radiation energy you absorb, related to your weight, i.e. how many joules of ionising radiation get absorbed by each kilogram of your body weight. In Martial Arts Land, a Gray measures how many kicks or punches actually landed on you — and how hard or soft they were. But here's a problem. If you suffer 50 little punches all over your well-muscled areas (e.g. arms and legs), you would not suffer any permanent damage. But if those 50 little punches were to your left eyeball, you might well be permanently blind.

That's why there is a third unit, the Sievert (Sv). It is roughly equivalent to a Gray, but has various weighting factors to work out how much damage you might suffer from the radiation (called the RBE or Relative Biological Effectiveness). For example, bone marrow is much more sensitive to radiation than your big toe. And one joule of alpha particles does 10–20 times as much damage as one joule of gamma rays, so the RBE for alpha particles is 10-20.

Radioactivity Units 202

An atom is extremely small, so the energy from the disintegration of a single nucleus is still small. Even so, when one atom of Uranium-235 disintegrates, it gives off enough energy to make a grain of sand visibly jump.

In real life, radiation physicists don't use the Becquerel (because it's small), so they use the gigaBecquerel (one billion Bq).

And, if they have to use the Becquerel, they humorously call it the 'Bugger-All' ...

References

Encyclopaedia Britannica, Ultimate Reference Suite DVD, 2006 — 'radioactivity'.

Ng, Kwan-Hoong, 'Radiation, mobile phones, base stations and your health', Malaysia Communications and Multimedia Commission, 2005.

CATS CAN SEE IN THE DARK

Cats have definitely wormed their way into our hearts as the most popular companion animal. They are so popular that they even have their own mythology, which includes the ridiculous claim that they can see in the dark.

History of Cats

Some 3000 years ago, in China and Japan, cats were held in high regard because they ate rats, and so protected the precious silkworm cocoons. The ancient Egyptians loved cats so much that they mummified them. Indeed, for a time during the reign of the Pharaohs, killing a cat meant the death penalty for the perpetrator.

In South Wales, during the 9th century, cats ate vermin and so protected the grain. For this reason a law was passed to stop people killing cats or their kittens.

Leonardo da Vinci obviously admired cats, noting that 'the smallest feline is a masterpiece'. And Thomas Huxley, the 19th-century British biologist, even claimed that the cat made the British Empire great by using some rather circuitous reasoning. Spinsters in the villages had cats, which ate mice. The bees (apparently the

I knew they were watching!

Cats cannot see in total darkness. They are, however, much better adapted than humans for seeing in low levels of light or semi-darkness.

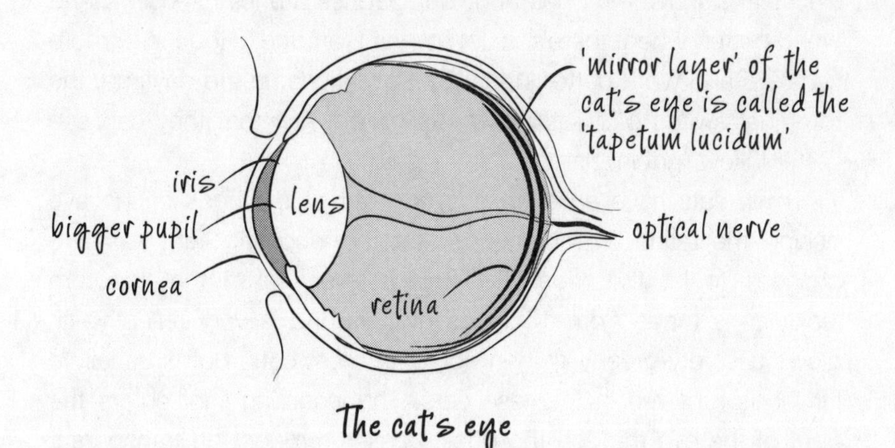

The cat's eye

Compared to the human eye, the cat eye can let in several times more light. The cat pupil (where the light comes in) is much larger than the human pupil.

favourite food of mice) then flourished and fertilised the red clover around the village helping it to grow well. The cattle then ate the clover and grew nice and meaty. And finally, loyal young British males ate the beef, which made them big and strong in order to man the ships that made the British Empire great.

Cats See Better in the Dark

The truth is that cats cannot see in total darkness any more than we can. Total dark is total dark. However, they are much better adapted than human beings for seeing in low levels of light or semi-darkness.

Cats use three clever evolutionary adaptations to allow them to do this.

First, compared to a human eye, the cat eye can let in several times more light. Compared to the size of the whole eyeball, the cat pupil (where the light enters) is larger than the human pupil.

Second, the cat eye is very richly endowed with rods. There are two types of cells in the eye that turn the incoming light into electrical signals — cones and rods. Cones are less sensitive and work better when there's a lot of light around. Rods are more sensitive and work better in low light situations. In the daytime, the rods just switch off. Cats have lots of rods, so that they can see better in low light levels.

Third, cats have an extra 'mirror' layer at the back of the eye behind the retina, which means that the incoming light has two chances to hit the rods. This mirror layer, called the 'tapetum lucidum' is made from 15 types of cells and is very reflective. It glows a silvery-greeny-golden colour in most cats, but is usually a luminous ruby-red in Siamese cats. The incoming light enters the front of the eye through the pupil, passes through the transparent innards of the eye and enters the retina. At this stage the light will register if it hits a rod. In the human eye, if the photon of light misses the rod, it is absorbed in a black layer behind the retina, and is gone forever.

But in the cat eye, if the light has not hit a rod, it will reflect off the mirror layer and bounce off. The light now has a second chance to hit a rod and be put to work. This effectively doubles the amount of useful light. You could say that the cat eye is a furry version of a military night vision system.

Cat vs Human Eye Reflection

When you shine a very bright light (e.g. a camera flash) into a human eye, you can't see the black layer because it's black.

But why you do see a reddish colour? The red colour comes from the red blood cells in the many blood vessels that nourish the retina. There are a lot of them, because the retina is a very oxygen-hungry organ. These blood vessels are in front of the retina, so they light up well.

Why don't we 'see' these blood vessels overlaid on our field of view when we look at the world? Because our brain 'electronically' snips them out of our vision. But the camera does see them, and hence the dreaded 'red eye' in so many flash photos. (By the way, you can sometimes see them when you close your eyelids and aim your face at a light.)

When you shine a bright light (such as a torch or car headlights) into the eyes of a cat at night, you see a reflection of this light, thanks to the 'mirror' in the back of the cat's eye.

Catseye Reflector

The concept of 'a mirror in the back of the eye' made Percy Shaw a very wealthy man.

In 1933, while working as a road repairer on the English roads on a foggy night, he saw this reflection in the eyes of a cat. He immediately came up with the idea of making artificial 'catseyes' to mark the centre of the road. He patented his catseye invention in 1934. It had a convex lens (like the lens in the human eye) in front of an aluminium mirror. This was set into a rubber pad, which in turn was encased in a cast-iron housing, which was then mounted in the road. It was aimed to reflect the headlights of a car back to the driver. In 1935, he began mass-producing them. In 1937, a British Ministry of Transport study found that his version of catseyes (or 'reflective road studs') survived better than those of his competitors. They are still used today.

So Percy Shaw, and his bank account, were able to bask in the reflected glory of his catseyes ...

Visual Electrophysiology 101

The cones (daylight vision) and rods (night-time vision) turn light into electricity.

The light enters from the front of the eye, landing on a cone or rod at the back of the eye. At the bottom end of a cone (or rod) is a section that also looks like a cone (or rod). It is packed full of parallel membranes covered with lots of unstable molecules called 11-cis-retinal. A photon of light carries an incredibly small amount of energy. Nevertheless, this energy is enough to make the molecule change shape from 11-cis-retinal to all-trans-retinal (from bent to straight). As it changes shape, it gives off a very small amount of electricity.

When lots of photons hit lots of molecules of 11-cis-retinal, they give off lots of electricity. This electricity is processed in the 10 or so layers of the retina, and then sent off to the brain. At the back of the brain are two areas that do extra processing on this electrical information, giving us this incredibly rich, wall-to-wall sensation that we call 'vision'.

Cat the Hunter

The cat is a superb hunter. The padded feet make no sound. Compared to its size, its muscles are enormous, making it very fearsome to any creature smaller than it — and to some creatures larger than it. The teeth have the ideal shape for ripping and shredding flesh. The claws are very sharp, and yet retract deeply so that they don't become blunt on hard surfaces.

The external ears have some 27 muscles to make them pivot and aim, in order to pick up sound better.

The cat has about 24 whiskers arranged in four rows on each side of the nose, and more whiskers above the eyes and on the front legs. The whiskers are very sensitive organs of touch. When a cat is about to deliver the 'killer bite' on its prey, the whiskers fan forward to feel the exact location of the prey, so that the bite can be more accurately positioned. And if it is totally dark a cat can navigate with its whiskers.

References

Did you Know?, London: Reader's Digest Association Limited, 1990, pp 38, 121, 124, 135, 136, 198, 263.

Exploring the Secrets of Nature, London: Reader's Digest Association Limited, 1994, pp 213, 323.

The Handy Science Answer Book, by The Science and Technology Department Carnegie Library of Pittsburgh, Detroit: Visible Ink Press, 1997, pp 307, 308.

The Origins of Everyday Things, London: Reader's Digest Association Limited, 1998, p 175.

The Reader's Digest Book of Facts, Sydney: Surry Hills, Reader's Digest Association (Australia) Pty Ltd, 1994, pp 17, 134, 309, 314, 320.

A DRIVE DOWN MEMORY LANE

During World War II, Alan Turing, the genius mathematician, helped crack the German military Enigma code. This gave the Allies a tremendous military advantage, because they could read Germany's coded communications. To do this, Alan Turing had to virtually invent the computer. He was one of the theoretical fathers of modern computers. However, he really got it wrong when he predicted that the total number of computers ever needed in the United Kingdom would be just six.

Today, most of us have a computer. In fact, we live in a world where we have many electronic 'toys' and don't really understand any of them. And so, many of us believe that when we 'erase' information from a computer's hard drive the information has really gone forever. If only this were true.

Love the Hard Drive

The modern hard drive is a cheap and wondrous device. My first 20 MB hard drive purchased in 1987 was the size of two phone books and whined like an aeroplane taking off in the distance. And it cost me $2500. At today's price of $1 per gigabyte, the same

20 MB of hard disc would cost only two cents. This is a lot less than $2500.

Inside the hard drive is a spinning disc (about 9 cm across) covered with iron ore (a fancy name for rust). Floating above it is a moving 'head' that can both read and write magnetic information into the rust.

To give you a mental picture of the relative sizes of all the parts, let's imagine all the sizes 200 000 times bigger.

The head is now the size of the Empire State Building (400 m x 100 m x 100 m) lying on its side. It's flying across the disc at a relative speed of 17 000 000 kph, 'floating' on a bubble of air just 2.5 cm thick. It's amazing that this distance is so small. The head (the Empire State Building) reads or writes a new bit of information (a '1' or a '0') every 10 cm, as it zooms along at 17 000 000 kph. The quality of the engineering needed to design and build this is astonishing.

Hard Drive Needs Air

The hard drive needs a bubble of air to keep the Empire State Building floating above the surface that it's almost touching. Hard drives are designed to run at ground level, where the air pressure is one atmosphere.

However, in an aeroplane cruising at altitude, the air pressure is about 80% of what it is at ground level. Each volume of air now contains 20% fewer air molecules, so it's easier to push through the thinner air. I remember one specific flight between Sydney and Perth. The cabin crew brought the meals around. I lifted up my laptop a little too quickly. In that instant, the Empire State Building pushed down through the 2.5 cm bubble of air, and hit the delicate surface of the spinning hard drive disc at 17 000 000 kph. Bits of rubble went flying everywhere. Red horizontal lines suddenly appeared on the screen and the laptop crashed. My hard drive was never the same again (RIP).

The world's largest astronomical observatory is situated on the top of Mauna Kea in Hawaii. Here the altitude is roughly 4200 m and, according to my little barometer watch, the air pressure is about 60% of ground level air pressure. The astronomers' hard drives kept failing, because the heads were more likely to bottom out in the thin air. The astronomers had to get fully sealed hard drives, filled with air at one atmosphere. Only then, could the astronomers use their hard drives reliably in the very thin air.

So try not to allow your computer to be bumped when you use it on a flight, where the atmosphere is 20% thinner and the bubble of air that holds up the Empire State Building is easier to push through.

Erase Hard Drive

Think of your hard drive as a simple filing cabinet. The filing cabinet has a drawer filled with 100 different hanging files, each labelled with a different name or number. And inside each hanging file is a single sheet of paper, which is also labelled.

Suppose you have 100 files on your computer. One of these files has the access details and passwords of every bank account that you have. This file, cunningly called 'Passwords', may take up several thousand 1s and 0s on your hard disc. One day, in a burst of paranoia, you 'erase' this file. But you have not really erased it. You have simply removed the name 'Password' from the list of files. You, the casual computer user, no longer know where your precious 1s and 0s are. But, and this is the important part, the 1s and 0s are still there.

You also have a sheet of paper labelled 'Passwords' in one of the hanging files in the filing cabinet. When you 'erase' the Passwords file, you don't really destroy the sheet of paper. All you do is hide its location (i.e. by removing the label on the hanging file). You will get rid of the sheet of paper only when you put another sheet of paper in the hanging file and remove and shred the old sheet of paper. Until you do that, the Password sheet of paper is still there.

In exactly the same way, your Passwords information is still on your hard drive until you write new information over it.

Are you sure you want to erase?

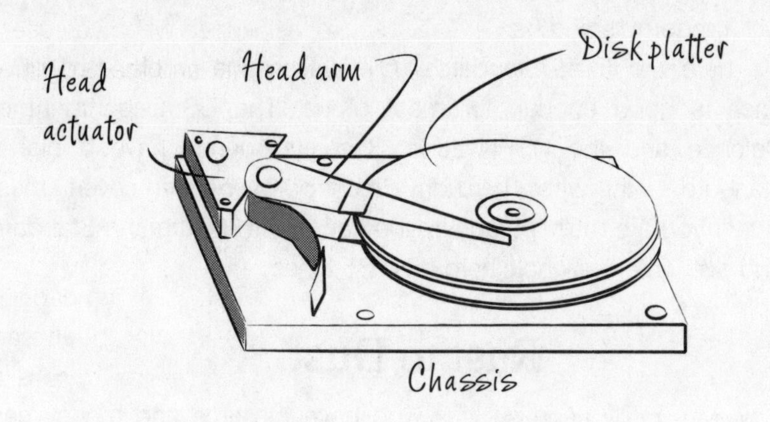

The typical internal of a hard drive

There is a common belief that once you 'erase' information from a computer's hard drive, the information is gone forever ... however, it ain't necessarily so. If you do a 7x or 35x overwrite with 1s and 0s, the info is truly gone. But a casual erase still leaves the info there.

Messy

All too often, government departments and companies around the world sell off unwanted hard drives loaded with sensitive information, e.g. employee details, legal agreements and access codes. These hard drives are either sold separately or still inside a computer. The New South Wales State Rail Authority did this early in 2005, but luckily the buyer did not misuse the information. Many other government bodies and companies have done this in the past, and will do so again in the future.

Sometimes, they foolishly don't even try to 'erase' the data. The next owner of the hard drive simply opens it up to find all kinds of sensitive information.

Mostly, people just do the standard 'erase' before selling the hard drive. But there are many software applications that will sniff out any information that is still on the hard drive.

The only way to really 'erase' information from your hard disc is to use a disc utility software that overwrites the whole hard drive with random 1s and 0s.

There are three standards. Overwriting the whole hard drive once is good enough for most of us. The US Department of Defence and the US National Security Agency have a higher standard — the whole hard drive must be overwritten seven times. But, if you are really paranoid, you can use the Gutmann Standard and overwrite the whole hard drive 35 times.

Rust to Dust ...

However, really keen snoops who have the time and money can use an atomic force microscope to sniff out traces of the original 1s and 0s. They are still there, even after the hard drive has been overwritten a few times with random 1s and 0s.

For this reason some military organisations go one step further to keep their information secret. They remove the actual iron ore–covered discs from their hard disc boxes and grind them into powder which they mix with cement to be used in the foundations of new buildings. If you want to crack open the concrete to find the microscopic bits of magnetised iron ore, go right ahead ... the secrets of the government could be yours.

Memory — Rust vs Rock

Many people are confused about the different types of memory in their computer. There is memory in the hard drive and memory in the RAM (Random Access Memory).

Two rules make it simple to understand. First, memory can be either rust or rock, and second, as Neil Young sang, 'rust never sleeps'.

The hard drive is made from purified rust. Rust is iron oxide (the red stuff that pushes through car paint if you live near the ocean). When you coat purified rust onto stiff metal plates and spin it at high speed, you can write magnetic information onto it. This is the hard drive memory. You can also read this magnetic information. 'Rust never sleeps' means that the information always remains on the hard drive, even after you switch it off.

The RAM is made from purified rock. Rock refers to silicon, which together with oxygen is a component of rocks. If you remove the oxygen and purify the silicon, you can make RAM memory, which normally sits inside the computer. It's fast but expensive. In general, install as much RAM for your computer as you can afford — it will run faster. The RAM is 'alive' only while the computer is switched on. So 'rock always sleeps', i.e. if you switch off the computer, the RAM goes blank.

Buy Hard Drives on eBay

In 2003, Simson Garfinkel, a privacy expert and graduate of the Massachusetts Institute of Technology, bought 158 second-hand hard drives on eBay. He collected over 5000 credit card numbers, financial and medical records, personal emails and lots of pornography. It was very easy, using readily available software.

Today, computer technology advances rapidly. A computer can become obsolete in just four years. What do you do with your old computers? What does your lawyer, doctor or accountant do with your personal data on their old computers?

Sell Hard Drives on eBay

The Idaho Power Company sells electricity to approximately 460 000 customers in Idaho and Oregon. In 2006, it hired Grant Korth, of Nampa, Idaho, to recycle 230 hard drives. He sold 84 of them on eBay. The drives had not been scrubbed and still contained confidential information on employees and customers, as well as internal and external correspondence.

Idaho Power Company has since decided that it will no longer sell old hard drives, but destroy them instead. Simson Garfinkel commented, 'The resale value of a hard drive is really minuscule. These things are worth US$5–$20 each. I don't think anyone's buying them on the secondary market for extortion, but you never know.'

Frances O'Brien, an analyst at Gartner Inc., said that it was common for hard drives to be sold while still carrying sensitive data. She said, 'It happens all the time. Typically, a user either doesn't know to clean the drives or doesn't do it correctly.'

Shrinking Hard Drive

Every year I read $10 000 worth of journals. And I used to store any relevant articles from the journals by photo-copying them. I was drowning in paper.

Nowadays, I capture the articles from the home pages of the various journals, and store them as PDFs (Portable Document Files, a format devised by Adobe). I chose PDF as a storage medium so that I would be reasonably sure that I could still read it in 10–20 years time. (Can you still read your old computer documents?) I add 1–2 GB of information to my hard drive each month — so I was running out of room on my laptop.

Laptop hard drives (2.5 in) are smaller than desktop computer hard drives (3.5 in). The limit for laptop hard drives used to be 100 GB. Then I found that a new Perpendicular Recording Technology had made 160 GB hard drives available. I ordered one from the USA for US$325 on 17 April 2006. Over the weekend, as my hard drive flew across the Pacific Ocean to Australia, the price dropped to US$289. By mid-June, it was down to US$269. I have just read that Hitachi will release a 200 GB laptop hard drive ...

References

Fisher, Sharon, 'Utility's disk drives — and date — sold on eBay', *Computerworld*, 8 May 8 2006.

Fitzgerald, Thomas J., 'Deleted but not gone', *The New York Times*, 3 November 3 2005.

McLean, Brad, 'Data erase', *Australian Doctor*, 13 June 2003, pp 53, 54.

Reagan, Brad, 'The digital detectives', *Popular Mechanics*, May 2006, pp 84-87, 134.

Tango, Jon William, 'Avoiding a data crunch', *Scientific American*, May 2000, pp 40-52.

MUSCLE TURNS INTO FAT

Some of us go to the gym and indulge in a strange activity called 'pumping iron' — or weightlifting. The 'pumping' comes from the lovely feeling you get after a good muscle-building session, and 'iron' comes from what the weights are usually made from. Many people worry that once they stop pumping iron regularly, their muscle will turn into fat.

But muscle cannot turn into fat, because muscle and fat have very different types of cells.

Cells 101

There are several hundred different types of cell in the human body — muscle, fat, liver, nerve, bone, skin, etc. It is actually a little more complicated, as each type of cell usually comes in a few different varieties. For example, there are three main types of muscle cells and two main types of fat cells.

Biologists define the cell as the basic unit of life — the smallest unit of the human body that is alive (and by the way, defining 'life' is pretty tricky). Each cell in the human body will grab and digest the food it needs, and then use this food to make energy, which in turn is used to make proteins. These proteins are either exported out of the cell into the body for general use (e.g. insulin and growth hormone) or else they are kept inside the cell to keep it alive and to make the next generation of cells.

A typical human cell is about 20 microns across (a micron is one-millionth of a metre), roughly one-quarter the thickness of a human hair. They can be smaller (e.g. red blood cells) or very long and skinny (e.g. the sciatic nerve, which runs over a metre from the lower spine to the toes).

Muscle Cells 101

Let's get specific about muscle cells. They come in three main types.

One type is smooth muscle, which is found in the walls of the gut and the uterus, as well as in the walls of some blood vessels. The second type is cardiac muscle, which is found in the heart.

Although both smooth muscle and cardiac muscle have their own internal pacemaker, they are usually driven by external factors. The smooth muscle in the gut is usually driven by the presence of food in the gut. The heart muscle by signals from blood pressure sensors in the neck.

However, the muscle worshipped by body builders is the third type — skeletal muscle — which is usually driven by signals from the brain. For example, your brain wants to read the next page of this book, and it sends a signal to various muscles in your arm to turn the page.

The individual muscle cells range from 30 cm long (sartorius muscle, which runs diagonally across the front of the thigh) to 1 mm (attached to tiny bones in the ear.)

If you dive deep enough into a muscle, you eventually end up at a structure called a 'sarcomere'. Inside the sarcomere are two types of microscopic filaments that lie next to each other like interlaced fingers. These are called 'thick' and 'thin' filaments. When an electrical signal arrives at the sarcomere, the thick and thin filaments slide over each other, shortening the sarcomere. When a whole lot of sarcomeres shorten, so does the muscle they belong to — and that's how a muscle works.

Fat Cells 101

There are two main types of fat cells — brown fat and white fat.

Brown fat can turn energy directly into heat. Because newborn babies cannot shiver, they have brown fat to keep them warm. Adults have very little (if any) brown fat. (Perhaps brown fat cells are involved in spontaneous human combustion — if it exists ...)

White fat cells have two main jobs. First, they store fats, which they grab from the bloodstream. Second, they make fats, from glucose and fatty acids, etc., which they also grab from the bloodstream.

However, fat cells do more than just store emergency food. They also insulate the body from cold weather and make many hormones, including leptin, which is involved in energy balance and the regulation of appetite. They are also involved in steroid metabolism. Curiously, deposits of fat around the belly are linked to insulin resistance, while fat on the limbs is more 'harmless'. In women, fat deposits on the hips, breasts and buttocks are somewhat controlled by female sex hormones.

When fully bloated (about 200 microns), white fat cells are filled with a large central blob of fat. This squashes all the internal machinery of the cell to the periphery of the cell.

Muscle vs Fat

As you can see, muscle cells are completely different from fat cells. Muscles do work, and get shorter as they do so. Fat cells have a completely different job, which is to store fat for when times are bad.

If you stop pumping iron, the muscles that you have so painfully built up get smaller. You still have the same number of muscle cells — but each muscle cell gets thinner. If you happen to keep eating the same amount of food, the fat cells (which were always there) will now expand. But, apart from the slow process of evolution, under no circumstance does one type of cell (muscle) turn into a completely different type of cell (fat) with different internal machinery, functions and shape.

The Muscle vs The Fat

There are three main types of muscle cells and two main types of fat cells found within our body.

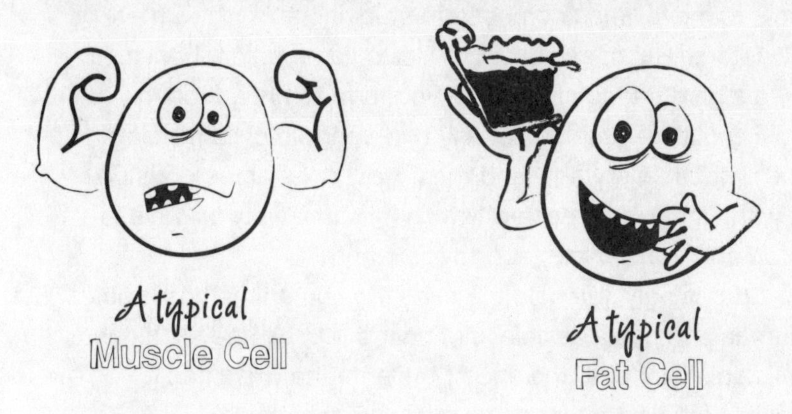

A typical **Muscle Cell**

A typical **Fat Cell**

Many folk believe that when you stop exercising, your muscle turns into fat. Muscle cannot 'turn into' fat because muscle and fat are two very different types of cells.

Fooling and Fat

We human beings have a magnificent capacity for fooling ourselves.

Dr S.W. Lichtman described this in the *New England Journal of Medicine* in 1992, in his paper 'Discrepancy between self-reported and actual caloric intake and exercise in obese patients'. The patients reported that they ate certain amounts of food (about 4000 kilojoules) and that they exercised (also equivalent to about 4000 kilojoules).

However, they actually ate an enormous 8400 kilojoules and burnt a modest 3000 kilojoules in exercise. If your 'kilojoules in' is bigger than your 'kilojoules out', then your fat cells will begin to bloat and gloat.

How an Octopus Eats

The vast majority of animals on Earth have skeletons (either internal or external). So most of our knowledge of how to move limbs comes from studying animals with skeletons. But in evolution, muscles came before bones.

It's hard enough to control your arm to bring food to your mouth — just look at any two-year-old child missing their mouth every now and then. How does an octopus do it with eight limbs, each of which has an infinity of ways to grab and transfer the food to the mouth?

Surprisingly, the octopus uses the human method. Each human arm has a flexible joint near the body (the shoulder), a flexible joint near the end of the arm (the wrist) and a flexible joint in between (the elbow). The octopus 'manufactures' three similar joints at similar locations, in its infinitely flexible arm. It 'makes' these joints by sending two waves of muscle contraction from each end of its arms. Where the waves hit each other, they stiffen the arm and turn it into a temporary joint. Each time it eats, it does this at three locations on each arm — at the same time!

We normally think of muscles as motor elements. But the octopus can also use muscles as structural elements.

Our gut runs some 10 m from our mouth to our anus. We still don't understand how this hollow muscle works. Perhaps we can learn more about our gut by studying the octopus more closely.

How a Tongue Works

When you send a nerve signal to a muscle, it contracts. When muscles work, they get shorter. You can bend your tongue in the middle like the letter 'L' and then push quite firmly against the roof of your mouth and the upper teeth on each side. So how does your tongue poke out, if muscles can only get shorter?

The way to solve this problem is to know that the tongue has many internal muscle fibres running in many different directions. You also need to think of the tongue as an organ that has a constant volume. When you poke your tongue out, you contract the muscle fibres that run across your tongue. The tongue has a constant volume, so it pushes forward.

References

Encyclopaedia Britannica, Ultimate Reference Suite DVD, 2006 — 'muscle'.

Encyclopaedia Britannica, Ultimate Reference Suite DVD, 2006 — 'adipose tissue'.

Lichtman, S.W., et al., 'Discrepancy between self-reported and actual caloric intake and exercise in obese subjects', *New England Journal of Medicine*, 31 December 1992, pp 1893-1898.

O'Connor, Anahad, 'The claim: muscle turns to fat when you stop working out', *The New York Times*, 26 July 2005.

Sumbre, G., et al., 'Octopuses use a human-like strategy to control precise point-to-point arm movements', *Current Biology*, 18 April 2006, pp 767-772.

FISH FEEL NO PAIN

Most of us have the vague impression that cold-blooded creatures such as fish don't feel any pain. This belief has been around for a long time. Only in the past few years have we probably proved that some fish do feel pain.

This is actually quite a controversial topic for some people, especially anglers who catch fish and members of animal rights groups. Tempers have flared, fists have been shaken, and much heat has been generated. Some people confuse the issue by claiming (with very little proof) that only creatures with a neocortex as part of their brain can feel pain. Under this strict definition, only human beings and primates can feel pain, and all other creatures, e.g. cows, dogs and fish, cannot.

Let's cut back on the heat and try to throw some light on this subject.

Pain — the Problem

The first difficulty is that fish cannot talk, and so cannot tell us if they are feeling pain. Another difficulty is that fish are not as cute as puppies, and don't get as much automatic and immediate sympathy.

There is another confusing issue — is the fish's reaction pain or reflex?

You won't feel a thing

Fish can't talk ... so it's hard to know if they are feeling any pain.

It's only in the last few years that it's been proved that some fish probably do feel pain.

For example, when you touch something hot, you pull your hand away. This does not happen because your brain feels pain, but because of a reflex. Pulling your hand away happens automatically thanks to your spine. Your brain is not involved. The heat activates the heat receptors in your hand, which send electrical signals to nerve centres in your spine. All by themselves, these nerve centres in your spinal cord activate the muscles needed to pull your hand away. This all happens without your brain being involved.

As a courtesy, these nerve centres in your spine also send an electrical signal up to your brain, to tell you what has just happened. A very small instant after you have pulled your hand away, you realise that your hand is feeling the pain associated with heat.

One part of your brain tells another part of your brain, 'Hey Karl, you just put your hand on the hot stove, again. Just thought that I would let you know why your arm suddenly pulled away. Do you mind not touching hot stuff again? And by the way, here's some pain to remind you not to do this again.'

When something bad happens to a fish and it reacts, does it react because of a reflex or because of pain? You can see the dilemma.

Pain — the Answers

The 'fish-pain' research was done by Drs Sneddon, Braithwaite and Gentle from the Roslin Institute in the UK (the home of Dolly, the cloned sheep) and the University of Edinburgh.

The scientists tested Rainbow Trout. Previous studies looked for pain receptors in fish with cartilaginous or non-bony skeletons (such as sharks or manta rays), and could not find any. But the researchers chose trout because they have a bony skeleton.

In a series of studies, they examined the fish for pain receptors, exposed them to heat, injected their lips with bee venom, saline solution or acetic acid, recorded brain activity and gave them morphine. They were able to show four results, three of which when combined, suggest that fish may feel pain.

Fish Have Pain Receptors

First, they showed that each trout had around 58 specialised receptors on its head. Some of these receptors responded to just one stimulus, while others responded to many.

Sixteen of them were some kind of pressure and touch receptor that responded slowly to physical pressure. Another 14 also seemed to be a type of pressure and touch receptor, but these responded quickly to physical pressure. Another six receptors responded only to mechanical and chemical stimulation, such as 1% acetic acid. Four receptors responded only to heat and mechanical pressure. Finally, 18 receptors responded to mechanical, heat *and* chemical stimulation.

In human beings, receptors that pick up heat and mechanical pressure are identified as being pain receptors. Therefore, in the Rainbow Trout, these last 22 receptors (4 + 18) could be counted as pain receptors. In fact, under the microscope, these receptors look virtually identical to the corresponding human receptors. They also had very similar mechanical and thermal thresholds. These receptors send signals to the fish brain via nerves.

So Rainbow Trout have the complete 'neuro-apparatus' needed to experience pain.

Fish React

Second, the lips of the fish were subjected to various nasty stimuli, while those of the control group of fish were not. The experimental group increased their heart rate by up to 30%. They also massively increased the rate at which they beat their gills. However, these reactions could be a simple reflex. It does not mean that their brains felt pain.

In general, you would think that it would be essential for an animal to be able to feel pain. The memory of the unpleasant sensation would make them keep away from immediate sources of injury. In the long run, this would help them live longer.

When you pull your hand away from the heat, it is because of a reflex, not because of the pain. The pain comes later. The pain occurs to teach you not to touch hot things again. The fish reacted to the nasty stimuli, but that's not enough to say whether they did or did not feel pain.

On its own, this 'reaction test' doesn't prove that the fish felt pain. But the next finding did suggest this.

Post-traumatic Stress

Third, the scientists observed that many of the fish had abnormal reactions after they had been injured.

Some of the fish adopted a 'rocking' behaviour, similar to what people who have had a close encounter with death will often do. (If you study TV footage of survivors of train accidents, earthquakes, etc., you will see some of them rocking back and forth, their arms clasped to their chest.) Some of the fish refused to feed for a long time after the injury. Other fish rubbed their lips in the gravel of their tank, something they would normally never do.

The fish suffered post-traumatic stress reactions. Some of these were almost identical to human post-traumatic stress reactions (e.g. rocking, not eating). Some reactions (e.g. rubbing the lips in the gravel) were the fish equivalent of us using our hands to rub sore lips.

Pain Killers

The last test was very interesting.

Dr Sneddon first first injected various nasty chemicals into the lips of the Rainbow Trout. As expected, they showed various kinds of abnormal behaviour and distress.

She then gave them a pain killer — morphine. The fish behaviour returned to normal.

Conclusion

The Rainbow Trout had all the necessary receptors and wiring to react to nasty stimuli. When hit with a nasty stimulus, they did what we would do — they pulled away and then showed abnormal behaviour. And finally, when given a pain killer their abnormal behaviour went away.

From these results, it seems fairly likely that fish (or Rainbow Trout at least) feel pain.

Hook, Line and Sinker

Dr Culum Brown, an ex-Queenslander currently at the University of Edinburgh, says, 'Fish are more intelligent than they appear. In many areas, such as memory, their cognitive powers match or exceed those of 'higher' vertebrates, including non-human primates.' And Dr Sylvia Earle, a leading marine biologist, said, 'Fish really are sensitive, they have personalities, they hurt when wounded.'

In our world, plants are about the only creatures that do not kill other creatures to get their raw materials and energy. Plants extract their raw materials from the air and the ground and their energy from sunlight. The rest of us have to eat other creatures, which involves the death of the creature, either plant or animal.

But we don't have to cause pain.

So what can you do the next time you catch fish? Take some ice and icy water with you and plunge the freshly caught fish into the icy water. As the fish cools, its metabolism will slow down, and it will sink (painlessly, we think) into hibernation and then anaesthesia. Then take it out of the water and place it gently on the ice. It will suffocate to death — without feeling a thing!

Problems with Sneddon Research

There are a few criticisms of the research done on the 20 Rainbow Trout.

First, '20' is not a very big sample size. I await a follow-up study with a larger sample.

Second, some people argue that we don't really know if fish have consciousness. The argument then goes that if they don't have consciousness, then they can't feel pain.

My understanding is that 'consciousness' is a very big philosophical problem that hasn't been solved for human beings, let alone fish. If you don't have 'consciousness', do you live in a mental darkness? Do fish have consciousness? Do dogs have consciousness? If the identical experiment were to be performed on dogs, could we say that the dogs experienced pain? Perhaps, feeling pain is different from understanding pain — but this doesn't block out the sensation of pain.

References

'Piscine pain', *New Scientist*, 6 September 2003, p 22.

Sneddon, Lynne U., 'The evidence for pain in fish: the use of morphine as an analgesic', *Applied Animal Behaviour Science*, 5 September 2003, pp 153-162.

Sneddon, Lynne U., et al., 'Do fishes have nociceptors? Evidence for the evolution of a vertebrate sensory system', Proceedings of the Royal Society of London, 30 April 2003, pp 1115-1121.

'Tender lips', *Science*, 9 May 2003, p 897.

A LITTLE ENLIGHTENMENT

The earliest known lamps are about 40 000 years old. Cave dwellers in Western Europe made lamps by hollowing out stones and burning animal fat in them through a wick made of moss. But lamps that burn fat or oil are a potential fire hazard — and create lots of soot and smoke.

That's why it was a great advance when Edison invented the electric light bulb ... except that Edison didn't actually invent it.

History of the Light Bulb

The modern light globe has a glass bulb filled with an unreactive gas, such as argon. The filament is made of tungsten, which resists high temperatures magnificently. Electricity is fed to the tungsten, which glows white-hot almost immediately. Thanks to the inert gas, the tungsten survives for approximately 2000 hours.

The light bulb could have been invented in 1666 — if somebody had seen the light. A giant storm hit London and for an hour or so lightning struck St Paul's Cathedral over and over again. The thick copper straps that carried the electricity down from the lightning rods on the roof to the ground glowed a dull red colour — just like an electric light bulb starting up. With this display, Londoners had seen electricity in action, but didn't realise what they were looking at.

A century and a half later, in 1801, Sir Humphrey Davy ran electricity through strips of platinum. They glowed but burnt out quickly when the platinum reacted with the oxygen in the air. In 1820, Warren De la Rue passed electricity through platinum inside a vacuum tube. The platinum survived, but it was too expensive and rare a metal for day-to-day use. In 1841, Frederick de Moleyns generated light by running electricity through powdered charcoal. He took out the first patent for an incandescent bulb. But again, his bulb didn't last long.

Oxygen Away

The secret to a long bulb life was keeping oxygen (in the air) away from the hot glowing material.

There are two main ways to do this.

The first way is to surround the glowing filament with a gas that will not react with it. In 1878, Hiram Maxim (inventor of the automatic machine gun) made a light bulb filled with petrol vapour instead of air — and the carbon filament inside worked perfectly well. (Yes, petrol vapour will burn if there's oxygen around. But if there's no oxygen it will not burn.) In fact, today's light bulbs have an unreactive gas (such as argon) inside, instead of a vacuum.

Second, you can remove all gases of all types from around the white-hot filament — in other words, surround it with a vacuum. (This needed the invention of good vacuum pumps.) In 1874, in Toronto in Canada, Henry Woodward and Matthew Evans patented a light bulb that had been emptied of air. The locals ridiculed them with, 'Who needs a glowing piece of metal?' In 1879, Thomas Alva Edison, realising that their patents relating to light bulbs were extremely valuable, bought them all.

Edison vs Swan

It was a close race between Sir Joseph Wilson Swan (England, 1878) and Edison (USA, 1879) for the first moderately reliable light bulb. They each came up with carbon filaments inside glass bulbs from which the air had been sucked out. Their early bulbs ran for a day or so before burning out. This was a vast improvement over the electric arc lamps of the day, which generated their light by making electricity jump a gap between two rods of metal or carbon. Not only did electric arc lamps give a very unpleasant light — and a bad smell — but they also needed continuous maintenance.

Edison and Swan sued each other for violating each other's patents. But being the bright sparks that they were, they eventually combined to form the Ediswan company, which became General Electric in 1892.

Who Invented the Light Bulb?

Edison did not invent the light bulb — he merely improved a 50-year-old concept. But he got all the credit for its invention.

Why? Because he was extremely clever, with 1093 patents to his credit — more than one for each month of his life. So it was automatically assumed that he 'invented' the light bulb.

Another reason was that he also developed all the essential technology needed for a practical lighting system. The concept of a 'system' embraced everything from the power plant (which included more efficient generators, as well as devices to monitor and control the current and voltage), circuit breakers, transmission lines, fuses and light sockets, right down to switches in the home or factory. He switched on his generating plant in Pearl Street, New York, close to the financial district, at 3.00 pm on 4 September 1882. His generator provided electricity to an area of about 2.5 km^2 of Manhattan, in which some 85 subscribers used about 400 of his light bulbs. The light bulbs burned out after 150 hours, but Edison improved that to 400 hours by 1884, and 1200 hours by 1886.

Again, Edison wasn't the first to build a fully functioning system. In 1881, a water wheel on the Wey River in Surrey in England had been linked to an electric turbine. The small system provided the very first supply of public electricity to a few houses and a few street lights in the small town of Godalming.

But Edison was clever enough to build his system not in a tiny hamlet, but in the financial district of one the biggest cities on Earth.

And he said ... 'Let there be light'

Edision definitely did NOT invent the light bulb. Instead, he improved a 50-year-old concept ... and got all the credit.

Coiled tungsten filament – this is the metal wire that glows when electricity flows through it.

A mixture of inert gases at low pressure.

The bulb is a thin layer of glass that surrounds the light bulb mechanism and the inert gases.

The common light bulb

The Very Smart Thomas Edison

He also had the financial backing to keep his system running until demand increased enough to provide a profit. In 1882, Edison's factories produced 100 000 light bulbs — and 45 million light bulbs in 1900.

The light bulb is a deceptively simple invention, but it has changed the way we live in so many ways.

How Light Bulbs Work

To understand how light bulbs give off light, you need to know that atoms have a central nucleus, surrounded by orbiting electrons. (Actually, under Quantum Mechanics, those electrons are also everywhere else in the Universe at the same time — but let's just ignore that for the time being.)

Electricity is applied to the tungsten filament which has resistance, and so heats up. Some of the heat energy appears at one particular atom of tungsten. The energy forces one of its many electrons to jump up into a higher orbit. When it falls back to its original orbit, it gives off a photon of light.

When this process happens to lots and lots of atoms, huge numbers of photons are given off and the light bulb filament glows.

Inspiration?

There's an old saying — 'Success is 1% inspiration and 99% perspiration'.

Edison and his team experimented with over 6000 different materials for the filament of the light bulb — including the beard hair of his assistants. He decided on charred bamboo threads in 1880, and tried a few thousand more materials before settling on tungsten.

Why Gas?

The tungsten filament gives off light perfectly well in a vacuum. But the tungsten is running white hot — not hot enough to turn it into a liquid, but very hot nevertheless. A small number of the atoms of tungsten get given enough energy to actually physically jump off the filament. Sometimes they land back on the filament, but usually at a different location. This leads to two problems.

First, the tungsten filament gets thinner at the locations where the tungsten atoms have jumped off. Because this increases the resistance at these locations, the filament gets hotter, making more tungsten atoms jump off. Eventually the filament gets so thin that it breaks, and the light bulb dies.

Second, the tungsten atoms often land on the relatively cool inner surface of the glass bulb. This causes the glass to darken, making the light bulb dimmer.

Adding an unreactive gas helps fix this problem. The gas (usually argon) circulates around the hot filament, helping to bring back many of the tungsten atoms to the very location that they jumped off from.

Longest Lived Light Bulb

A light bulb made in 1901 is still burning in a fire station in Livermore, California.

Dennis Bernal, who owned the Livermore Power and Light Company donated it to the fire station in 1901. It has burnt continuously since then, apart from two changes of location, and a few short, unplanned blackouts

References

Binney, Ruth (Ed.), *The Origins of Everyday Things*, London: Reader's Digest Association, 1999, pp 26, 27, 152, 153, 177, 223, 250.

Brain, Marshall, *Marshall Brain's More How Stuff Works*, Milton, Queensland: Wiley Publishing Australia, 2003, p 140.

Panati, Charles, *The Extraordinary Origins of Everyday Things*, New York: Harper & Row, 1987, p 136.

MYTHS OF GLADIATORIAL PROPORTIONS

For most people, the word 'gladiator' brings to mind either Russell Crowe (in *Gladiator*) or Kirk Douglas (in *Spartacus*). For this reason, most of us believe that gladiators were escaped prisoners of war or criminals, who were on the road continuously fighting in wild free-for-all melees. And of course all bouts finished with at least one gladiator dying. Like all successful myths, these beliefs have some nuggets of truth.

History of Gladiators

The word 'gladiator' means 'swordsman' – from the Latin *gladius* which means 'sword'.

In the gladiator schools (*ludi*), professional instructors (*doctores*) trained the students, beginners starting with a wooden sword (*rudis*). The seven main classes of gladiators were distinguished by the weapons they used or their method of fighting. The Retiarius (net man) wore a short apron, padding on the right arm, a protective mini-shield on the right shoulder and a throwing net in the right hand. The Secutor and Murmillo had similar weapons — a shield in the left hand, a padded left shin and right arm, and a long knife in the right hand. The Samnites fought with a large shield, a visor, a plumed helmet and a short sword, while the Thracians had a small round buckler (shield) and a dagger curved like a scythe.

Originally, the gladiators performed to the death at Etruscan funerals. The ones who died had the noble role of being bodyguards for the dead man in the next world.

Roman promoters recognised an opportunity, changing the emphasis from providing funeral attendants to entertainment. Gladiator bouts soon became very popular in Rome. The first documented gladiator bout in 264 BC had just three pairs of gladiators. By the time of Julius Caesar (who died in 44 BC), 300 pairs of gladiators entertained the crowd. The Emperor Titus stretched the shows from just one day to 100 days. In 107 AD, the Emperor Trajan celebrated a triumph with 5000 pairs of gladiators

'At my signal, unleash hell.'

Most people believe that gladiators were escaped prisoners of war or criminals who were constantly fighting for their lives. However, they were highly trained fighters who battled 2–3 times a year ... with a 1 in 9 death rate.

in the arena. Fifty thousand people would crowd into the Colosseum to enjoy the sport. The spectators could even buy special souvenir glass cups embossed with the names of their favourite gladiators. Gladiatorial bouts came to an end early in the 5th century, probably because of Christian opposition and the great costs involved.

So, What About the Myths?

Slaves?

Yes, many gladiators were slaves or condemned men, but some were free.

The free men were attracted by the money and the lifestyle (e.g. the favours of society women) and were not particularly worried by the one-in-nine death rate. Some men who had been financially ruined saw this as a way of rehabilitating themselves.

Continually Fighting?

It would have been too exhausting for the gladiators to fight continually, week in and week out. They would have sustained too many injuries to be able to compete. Instead, on average, they fought two or three times each year.

If they were slaves, and if they survived long enough — 3–5 years — they could sometimes earn their freedom.

Wild Melee or Choreographed Wrestling?

The bouts were definitely not undisciplined free-for-alls. The main purpose was to provide entertainment for the audience (compare this to the precise choreography in modern TV wrestling).

The gladiators were carefully matched in pairs from the seven categories, taking into account the attack and defence weapons they carried and the strength and skill of each individual. So the Retiarius would usually be matched with the Secutor (pursuer), but sometimes with the murmillo.

Steve Tuck, an archaeologist at the University of Miami, examined 158 images of gladiatorial art to try to better understand their fights. Gladiatorial art was quite popular, and these 158 images came from large-scale wall paintings, gems, cheap Roman lamps, etc. He then compared these 2000-year-old Roman images to more recent images from fighting manuals made in medieval and Renaissance times in northern Italy and Germany. These later manuals showed sequences of moves accompanied by instructions.

In both the Roman and later images, there were three major stages in any fight. The first stage was the initial contact, where the fighters came towards each other and tried to land a good blow. The second stage had an injured fighter, trying to keep his distance from the other fighter. The third stage had both fighters dropping their weapons to the ground, before wrestling with each other.

Gladiator bouts were like a sophisticated entertainment version of martial arts. They were closer to modern choreographed TV wrestling than wild melees. Like modern Western boxers who follow the Marquis of Queensbury rules, gladiators had their own very strict (and definitely more brutal) rules, which two referees enforced.

Drs Klaus Grosschmidt and Fabian Kanz examined a 1800-year-old gladiator graveyard in Ephesus (Turkey). They found that each skeleton showed a single characteristic pattern of injuries, supporting other reports that each gladiator usually fought in one single style, usually against another specific style of gladiator.

To the Death?

Yes, of course deaths happened in gladiatorial battles — but not every time.

It was simply too expensive to finish every single bout with one of the two highly trained gladiators dying. It took lots of time, training, hard work and money to get gladiators skilled enough to entertain 50 000 spectators. There are many references to the gladiators being trained to subdue, not kill, their opponent. The bout had to end in a decisive outcome, such as defeat through death (rare) or defeat through injuries or exhaustion.

If one of the gladiators was too badly injured to survive, he would be killed cleanly backstage after the event, with a single hammer blow to the side of the head. And if the crowd had condemned the gladiator to death and he had not died in the arena, he would be dragged backstage to be killed. Even though the gladiators usually wore helmets, 11 of the 67 skulls in the Ephesian graveyard showed this specific hammer injury. Importantly, the hammer was not one of the weapons used in the arena.

Finally, the thumbs-down gesture usually meant that the gladiator was already dead, not that he should be put to death. It was usually made by the organiser to the spectators.

Spartacus and Gladiatorial War

After Spartacus had served in the Roman Army, he became a bandit and was caught and sold as a slave. In 73 BC he escaped from his gladiatorial training school at Capua with 70 other gladiators and started a rebellion against the Romans. At his peak, he led 90 000 rebels against the Roman Empire, won battles against Roman consuls and then tried to leave Italy.

But his soldiers wanted to stay. In 71 BC, his army lost a battle against the eight legions of Crassus, in which Spartacus died. Crassus then crucified 6000 of the survivors on the Appian Way.

First Sword Fight

The very first illustration of a sword fight dates back more than 3000 years to 1190 BC and is shown in a relief in the temple of Medinat Habu, near Luxor in Egypt.

The image probably depicts a practice fight, because the swordsmen are wearing lots of protective padding (which would slow them down in a real fight) and the points of their swords are covered.

References

Encyclopædia Britannica, Ultimate Reference Suite DVD, 2006 — 'fencing'.

Encyclopædia Britannica, Ultimate Reference Suite DVD, 2006 — gladiator'.

Kanz, Fabian and Grossschmidt, Karl, 'Head injuries of Roman gladiators', *Forensic Science International*, 15 November 2005, pp 1-10.

Marks, Paul, 'Gladiators fought by the book', *New Scientist*, 23 February 2006, p 17.

Radowitz, John von, 'Fair fight: research showed that gladiators played by the rules', *The Sydney Morning Herald*, 24 February 2006, p 9.

Young, Emma, 'Gladiators fought for thrills, not kills', *New Scientist,* 22 January 2005.

HARDWOOD NOT HARD

I am lucky enough to have a chippie (carpenter) for a brother-in-law, and over the years he has taught me a lot. I don't know much about carpentry but I can do one or two things with wood — screw lumps of wood together, or drill holes and bolt lumps of wood together. However, I was pretty sure about one thing — hardwood must be harder than softwood. But I was wrong.

I was surprised to discover that hardwood can be soft and softwood can be hard. By 'soft' I mean that the material will dent or mark easily if you press it with a sharp point. And 'hard' means that you have to press the material a lot harder to make a similar dent or mark.

Tree Classification 101

It's always difficult to find an accurate way of classifying complicated, living creatures. Trying to classify trees into 'hardwood' and 'softwood' is a good example of this.

Softwoods are (in general) older and more 'primitive' trees, i.e. they evolved first. They usually keep their leaves year round. Hardwoods are (in general) more recent trees (speaking from an evolutionary viewpoint). They are flowering plants that produce seeds in a ripening fruit, e.g. a gumnut or an apple. Some hardwoods lose their leaves and sleep through the winter.

The hard (and soft) facts

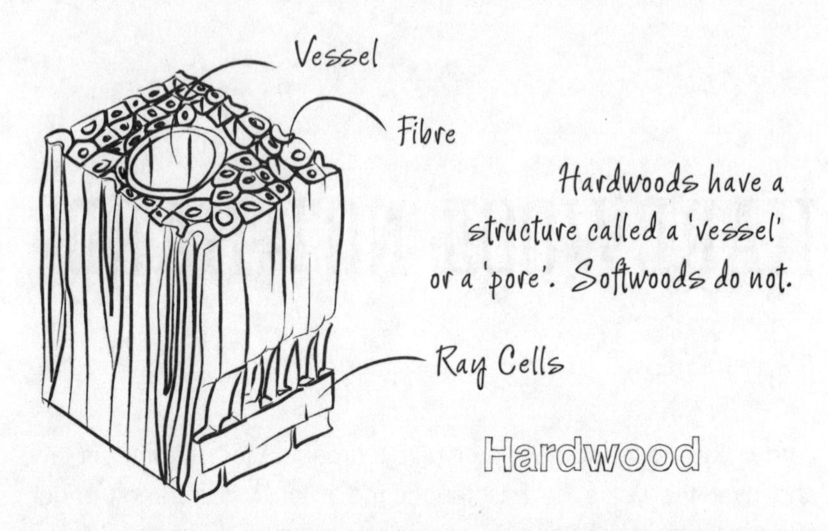

Softwoods are (in general) older and more 'primative' trees (meaning they evolved first). Hardwoods are (in general) more recent (speaking from an evolutionary point of view).

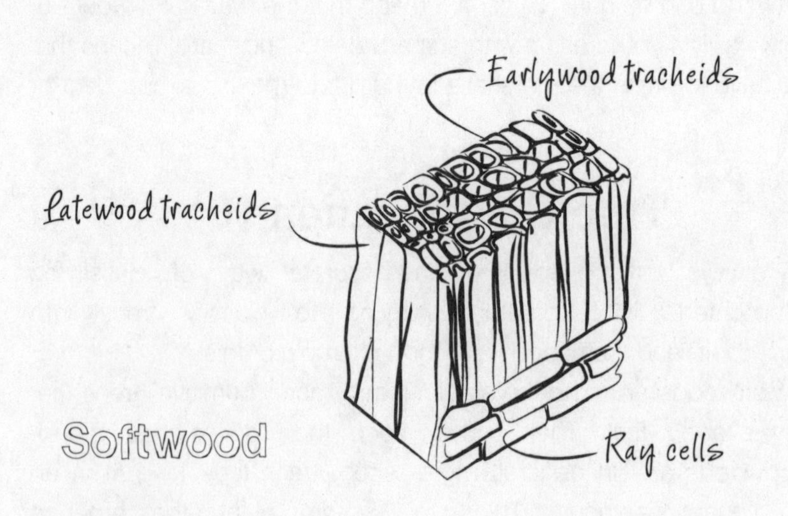

Two major differences between these trees are 'sex' and 'cell tissue'.

Sex

Softwood trees (gymnosperms) have sex in a more 'primitive' way. Their sex organs are on little cones, the wind blowing the pollen from one softwood tree to another. This is so striking in some American forests that everything downwind turns yellow from the pollen. The foresters call this time of year 'when the pollen flies'. These cone-like structures give the trees their name — conifers. There are about 600 different species of conifers, commonly found in cool and temperate climates. Conifers don't have flowers.

On the other hand, in the vast majority of hardwood trees (angiosperms), the sex life usually involves insects (or birds or other creatures) carrying the pollen from one flower on a tree to another. In a few species of hardwood (hazelnut, willow, beech, etc.) the wind carries the pollen. There are many thousands of hardwood species and they live in every climatic zone.

Cell Tissue

The other major difference between softwoods and hardwoods is their cell tissue. Hardwoods have a structure called a 'vessel' or a 'pore', while softwoods do not. Using a simple, low-power hand lens (x10 or x20) you can see the cross-section of a vessel if you make a clean cut to the end of a hardwood timber with a sharp blade. Because 'vessels' are also called 'pores', hardwood is often called 'pored' timber in the timber trade. Softwood is often called 'non-pored' timber.

Today, about 20% of the lumber produced worldwide is hardwood (e.g. eucalypt, teak, oak, American black walnut, maple, willow, poplar and beech). The remaining 80% is softwood (e.g. longleaf pine, Douglas fir, cedar and yew).

Hardwood Can Be Soft

What really surprised me is that the names 'hardwood' and 'softwood' have nothing to do with whether the wood is hard or soft. The hardwood called balsa is one of the softest woods known. And Australian Cypress (which botanists classify as a softwood) is quite hard.

How did the names arise? The answer unfortunately, is lost in antiquity, but hardwoods probably got their name from early carpenters working with hard European woods such as oak and beech.

What is Wood?

There are three major chemical components of wood.

'Cellulose' makes up 40–45% of wood. It's actually thousands of glucose molecules joined end to end to make long chains.

'Hemicellulose' makes up about 20% of softwoods and 15–30% of hardwoods. The actual chemical structure of hemicellulose is different in softwoods and hardwoods. It's a gelatinous material, in which the cellulose chains are held in a matrix.

Finally, 'lignin' makes up 22–30% of wood, the percentage slightly higher in softwood than in hardwood.

References

Bootle, Keith R., *Wood in Australia: Types, Properties and Uses*, Sydney: McGraw-Hill, 1985, pp 3-10.

Encyclopaedia Britannica, Ultimate Reference Suite DVD, 2006 (accessed 12 June 2006) — 'tree'.

ROACHES AND RADIATION

You've probably heard it said that come the end of the world, the only survivors will be the cockroaches. This 'fact' has struck a chord with the public and has been accepted into modern-day folklore.

Cockroaches have been around for about 280 million years, outlasting the dinosaurs by about 120 million years. They are very tough little critters that can survive on cellulose and, in a pinch, each other (yup, they're cannibals). They can even soldier on without a head for a week or two — and they're fiendishly fast as well. They have a reputation for being survivors, living through anything from steaming hot water to a nuclear holocaust. But at this stage we should start to be sceptical. Cockroaches are only a bit better at surviving radiation than we are, and are well and truly outranked in the nuclear holocaust survivor stakes by many other creatures.

Radiation and Insects

In 1919, Dr W.P. Davey became one of the first people to test how well insects survived radiation when he sprayed the Flour Beetle with small doses of x-rays. He was astonished to find that a dose of 60 rads seemed to make the Flour Beetle live longer. Surprisingly,

Dr J.M. Cork found the same result when he repeated the experiment in 1957.

A more typical result (of radiation harming living creatures) was found by Dr H.J. Muller in 1927, when he used x-rays to cause mutations in the fruit fly. (Even today, there is debate over whether low doses of radiation can be helpful for your health! This is called 'radiation hormesis'.)

Radiation and People

Until the late 1940s and 1950s, there really wasn't a lot of research done on being able to survive radiation. Around this time, three factors came into the equation.

First, there were now people who had been exposed to a lot of radiation and survived — the victims of the two atom bombs dropped on Japan. Second, there was the start of the Cold War and the nuclear standoff between the Superpowers. (At the peak of the Cold War, there were about 50 000 nuclear weapons on the planet.) Finally, there was the search for peaceful uses of nuclear power.

As a result, we discovered that human beings are much more susceptible to radiation than insects, and will die after a dose of 400–1000 rads. People as far as 21 km from Ground Zero at Hiroshima received doses of 1200 rads, suffering slow and agonising deaths.

Insects Rule

It turns out that insects are much more resistant to radiation than human beings. Wood-boring insects and their eggs were able to survive doses of 48 000–68 000 rads with no apparent ill effect. In 1959, Drs Wharton and Wharton found that it took 64 000 rads to kill the fruit fly (*Drosophila melanogaster*), and a colossal 180 000 rads to be sure of killing the parasitoid wasp of the genus *Habrobracon*.

As a result of all this testing, it gradually emerged that the cockroach is, at least in terms of nuclear survivability, a wimp. In 1957, the two Drs Wharton had found that it took only 1000 rads to interfere with cockroach fertility. In 1963, Drs Ross and Cochran

You dirty (but resilient) little 'roach ...

Cockroaches have been around for about 280 million years. BUT ... cockroaches are only marginally better at surviving radiation than us humans.

found that a dose as low as 6400 rads would kill 93% of immature German cockroaches — making cockroaches only 6-15 times tougher than frail human beings. All of them died at 9600 rads.

Cockroaches survive radiation about 10 times better than we do, but curl up and die at doses than don't even bother other insects.

How then did cockroaches get this reputation of being indestructible? Perhaps a cockroach looks more the part of a mean, radiation-resistant insectoid villain than a fruit fly does. If cockroaches were close to Ground Zero of a smallish 15-kiloton Hiroshima-class nuke, they would die. Cockroaches certainly could not survive the larger megaton-range hydrogen bombs in today's nuclear stockpiles.

King of Radiation

At the moment, the real King of Radiation is a foul-smelling reddish bacterium called *Deinococcus radiodurans* (known as Conan the Bacterium by its admiring researchers). It was discovered in the

1950s growing happily in canned meat that had gone bad, even though the meat had been sprayed with radiation to preserve it — a nice example of evolution. This bacterium frolics happily in background levels of 1.5 million rads of radiation, and seems to be able to survive twice as much radiation when in a frozen state.

Knowing that cockroaches are almost as vulnerable to nuclear attack as the rest of us doesn't make them seem any more lovable to me.

Papa Roach

The metal/punk band Papa Roach chose the name 'because a cockroach can survive anything: earthquake or nuclear holocaust'. Perhaps Papa Roach should change its name to Stinky Pinky (after the foul-smelling, reddish Conan the Bacterium) ...

Radiodurans

Conan the Bacterium has one major trick to help it survive massive radiation levels — redundancy. A classic example of redundancy is using both a belt and braces to hold up your trousers.

In the case of *Deinococcus radiodurans*, the DNA (its genetic code or blueprint) has massive redundancy. There are lots and lots of copies of the same section of DNA. If one section is damaged, it is compared to identical copies in other areas, and then repaired.

Scientists have been genetically modifying it to be able to 'eat' radioactive wastes, so as to remove them from the general environment.

Cockie Myth — Chromosomes

A long time ago, someone told me that cockroaches survive high radiation because they have big chromosomes. This means that the incoming radiation is proportionately smaller. Radiation hitting a small chromosome is like a basketball hitting a tennis ball — the tennis ball is pushed aside. Radiation hitting a big chromosome is like a basketball hitting a 50-tonne army tank — the tank doesn't move and the basketball bounces back.

It makes sense — except that it's totally wrong. Cockroaches don't survive high doses of radiation very well, and they don't have large chromosomes.

Cockroach DNA is roughly the same size as human DNA. However, cockroaches have more chromosomes in each cell (about 60, instead of 46). So each of their chromosomes is, in fact, smaller.

Cockie Nationality Myth

The German cockroach *Blattella germania* (the small, light brown cockies that you find around the house), is actually from Asia. They moult every seven days and become adults at about 60 days of age.

The American Cockroach *Periplaneta americana* (the large, dark cockies that you find arrogantly strolling the streets), comes from Africa. They moult every month, become adults after six months and live up to two years of age.

Cockroach Control

The very first remedy for controlling cockroaches dates back to the 18th Dynasty in Egypt (1750–1304 BC). It was a curse uttered by Khnum, the ram-headed god — 'Be far from me, o vile cockroach, for I am the God Khnum'.

References

Berenbaum, May, 'Rad roaches', *American Entomologist*, Fall 2001, pp 132-133.

Daly, M.J., Venter, J.C., Fraser, C.M., et al, 'Genome sequence of the radioresistant bacterium Deinococcus radiodurans R1', *Science*, 19 November 1999, pp 1571-1577.

Gordon, David George, *The Compleat Cockroach*, Berkeley, California: Ten Speed Press, 1996, pp 149-170.

Ross, M.H. and Cochran D.G., 'Some early effects of ionizing radiation on the German cockroach, Blattella germanica', *Annals of the Entomological Society of America*, 1963, 56: 3, pp 256-261.

DADDY LONG LEGS

My seven-year-old daughter Lola shared a secret with me. 'Did you know, Daddy', she said very solemnly, 'Daddy Long Legs spiders are the most poisonous spiders in the world, but they can't hurt us because their fangs are too short to get through our skin.' She knew this, because she heard it from her seven-year-old friend India, who heard it from Charlie, India's 10-year-old sister — so it had to be right. But the children were wrong on four counts, which has to be some kind of record for one sentence.

This myth is known around the world, not just Australia. Most people know the Daddy Long Legs as a spider-like creature with a tiny body and long spindly legs, spanning roughly the size of a 20-cent coin.

Spider?

First, not all Daddy Long Legs are spiders — some are and some aren't.

The Class Arachnida is divided into several Orders — Order Araneae (spiders), Order Scorpiones (scorpions), Order Acari (ticks and mites) and Order Opiliones (opilionids).

Some Daddy Long Legs are opilionids. In this case, they are also called long-legged harvestmen. At a casual glance, these opilionids look a little like spiders — but they are not (because they do not belong to the Order Araneae). Opilionids have two eyes (spiders have eight), one body section (spiders have two) and an abdomen that is clearly separated into segments (spiders' abdomens are unsegmented). They do not make silk and have a very different respiratory system from spiders.

However, in the Order Araneae there is one real spider that looks like your classic Daddy Long Legs, so it is often called by this name. Its scientific name is *Pholcus phalangoides*.

Daddy Long Legs found inside your house are probably spiders. But those found in dark damp places like the shed are probably harvestmen.

Poisonous?

Second, are they 'poisonous'?

Actually, this is a little pedantic. Strictly speaking, 'poisonous' means that the creature exudes poison from its skin or shell. If the poison lands on your skin, it will continue its dirty work. When talking about a creature that injects poison through your skin into the soft layers underneath, the correct word is 'venomous'.

But what the heck, the English language is continually changing and I can let the kids have this one.

Most Poisonous or Venomous?

Third, are they the 'most poisonous' or the 'most venomous'?

Opilionids do not have poison glands. Nor do they have fangs through which to squirt their nonexistent poison.

There are about 20 000 different species of spider on Earth, but only about 50 of them are venomous. The American Black Widow and the Australian Funnel Web are heavy hitters. Because the spider *Pholcus phlangoides* is a lightweight, it took many years for anybody to bother to extract the toxin. You do this (as seen on the TV show *Mythbusters*) by anaesthetising the spider with carbon dioxide and then delicately applying tiny electric shocks to the right part of the spider. By then placing a very thin glass pipe up against the fangs, you can collect about two-billionths of a litre of toxin per

These legs are made for walking ...

The Opilionid – the 'spider' you have when you don't have a spider.

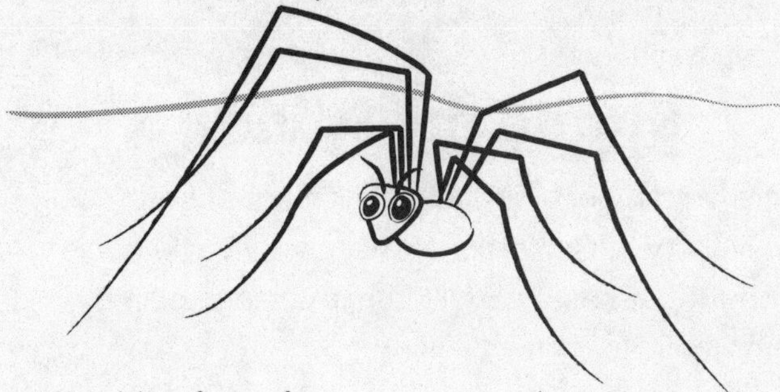

Some Daddy Long Legs are opilionids – they look like spiders, but they are not. Opilionids have two eyes (not eight like spiders) and one body section (spiders have two).

spider. When you do this a few hundred times, you get enough toxin to test. However, it's incredibly weak, barely worthy of being counted as a poison.

Short Fangs?

Fourth, are the fangs too short to penetrate your skin?

Once again, the TV show *Mythbusters* actually measured the fangs on one spider (a small sample size, but it's a start). Their spider wrangler, Chuck Kristensen, found that the fangs were 0.25 mm long. But skin, especially on non-contact areas, can be as thin as 0.10 mm.

A Rare Bite

A *Mythbusters* host, Adam Savage (who is terrified of spiders), bravely put his entire arm into a perspex box filled with what looked like hundreds of the spider version of Daddy Long Legs. After what seemed an eternity, one bit him — and he lived, complaining only of a mild sting.

So why the bad rap for the Daddy Long Legs? Spiders have some very hairy image problems. Perhaps they should cut back on the web spinning and go all out on the spin doctoring …

Blame the Spider

Dr Richard Vetter works in the Department of Entomology at the University of California at Riverside. Some of his studies show just how wrongly human beings blame spiders for their ills.

One of his studies focused on 106 people with the disease called 'lymphomatoid papulosis', a rare, low-grade, cutaneous T-cell lymphoma. It causes skin eruptions, which then heal spontaneously over a few weeks. This non-fatal disease is definitely not caused by any type of bite. He looked at 106 people with this disease.

Of the 106 people interviewed, 16 were told that they had been bitten by a spider. Of the 16 who were wrongly diagnosed with spider bites, five were told that they had been bitten by the Brown Recluse spider. But four of those five lived in areas where the Brown Recluse spider did not exist.

The world is definitely 'spiderist' …

References

Isbister, Geoffrey, 'Necrotic Arachnoidism: the mythology of a modern plague', *The Lancet*, 7 August 2004, pp 549-553.

Vetter, Richard, 'Myths about spider envenomations and necrotic skin lesions', *The Lancet*, 7 August 2004, pp 484,485.

MOUSE WITH A HUMAN EAR

In 1997, a rather bizarre photograph suddenly became very famous. It showed a totally hairless mouse with what appeared to be a human ear growing on its back. This photograph was used to start a wave of protest against genetic engineering which continues today. But there was absolutely no genetic engineering involved in getting the ear on the mouse's back.

Spare Ear?

Why would anyone need a 'spare' human ear? Because plastic surgeons have to repair ears quite often, and it's a very difficult job. The external ear — the bit you hold your sunglasses up with — is often damaged in car accidents, fights or fires. Also, it sometimes needs to be repaired in the condition known as 'microtia', which literally means 'small ear'. In microtia, the ear can range from being slightly smaller than average to being almost completely absent. Depending on the society, microtia can occur in up to one in 1000 births.

The ear is mainly cartilage, which is tricky to work with because its properties are a peculiar mix of bone and skin. At the same time, the ear has a highly visible and complicated shape. So a spare ear would solve a lot of problems. The famous Indian surgeon Sushruta, who lived somewhere between 1000 BC and 100 AD, described operations to repair ears. He wrote that the surgeon should 'slice off a patch of living flesh from the cheek of a person devoid of ear lobes in a manner so as to have one of its ends attached to its former seat'. Such surgery did not reach Europe until the second half of the 15th century.

Even today, surgeons have problems repairing the ear and would love to have a 'spare' ear to transplant.

The Famous Photo

In August 1997, Joseph Vacanti (a paediatric surgeon) and his colleagues wrote their ground-breaking paper in the journal *Plastic and Reconstructive Surgery*. The publicity was enormous, helped by a film made by the BBC's *Tomorrow's World*. It was a BBC photographer who took the famous photo.

On 11 October 1999, the anti-genetics 'Deep Environment' group, Turning Point Project, placed a full-page ad in *The New York Times*. It showed the photo with a misleading caption that read, 'This is an actual photo of a genetically engineered mouse with a human ear on its back'. The caption was inaccurate. First, the mouse had not been genetically engineered. Second, the 'human ear' had absolutely no human cells in it.

Genetically Engineered?

What would you have to do to get a 'genetically engineered mouse', as the inaccurate Turning Point advertisement described it?

You would have to alter the mouse's DNA — its genetic 'blueprint'. The first step would be to find a very specific chunk of human DNA — the section that has the blueprint for making the human ear. The next step would be to remove this section from the human DNA and insert it into the mouse DNA. Finally, you would have to command this foreign human DNA to wake up and grow a human ear — and on the back of a mouse, not anywhere else.

Mouse ears

In 1997, a photograph of a mouse with a human ear growing on its back became famous.

This didn't mean a mouse grew human ears that replaced its own.

There was a common belief that it was genetically engineered, but this was far from the truth.

But *none* of this happened. The mouse had not been genetically engineered. This process is beyond the technology available in 2007, let alone in 1997 when the photo was taken.

The actual process used was quite different. It's called Tissue Engineering.

How to Grow an Ear-like Object

The 'mouse-ear' project began in 1989, when Charles Vacanti (brother of Joseph) managed to grow a small piece of human cartilage on a biodegradable scaffold. His plastic surgeon colleagues had told him that the human ear was the body's most difficult cartilaginous tissue to reconstruct and rebuild, and that they would love to have a 'spare' ear to transplant. (Charles was then Director of the Center for Tissue Engineering at the University of Massachusetts Medical Center, and is currently Professor of Anaesthesia at Harvard Medical School.)

The team chose a three-part recipe: first, make a biodegradable scaffold; second, seed it with cartilage cells; and finally, let it grow.

The scaffold was the synthetic material (99% polyglycolic acid and 1% polylactic acid) used to make surgical stitches that dissolve. In the body it degrades away completely into carbon dioxide and water. The fibres of this material were woven into a loose mesh that was 97% air, leaving lots of room for cells to grow into. Even this first step was difficult. It took Charles's team eight years before they could mould their sterile biodegradable mesh into the exact shape of a three-year-old child's ear.

The next step was to seed this ear-shaped scaffold with cartilage cells from the knee of a cow. The famous mouse-ear had absolutely no human cartilage cells in it.

For the third step, the team used a Nude Mouse. The Nude Mouse got its name thanks to a random mutation in the 1960s that left the mouse with no hair and virtually no immune system. The lack of hair was irrelevant to their project, but the lack of immune system was critical. Having no immune system, the mouse would not reject the foreign cow-cartilage cells. (Nude mice have so few rejection issues that they have even grown transplanted feathers!)

The only purpose of the mouse in this project was to supply 'power' to help the cow-cartilage cells grow. The synthetic ear seeded with cartilage cells was transplanted onto the mouse's back — under the skin layer but over the muscle layer. Over about three months, the mouse grew extra blood vessels that nourished the cow-cartilage cells, which then grew and infiltrated into the biodegradable scaffolding (which had the shape of a human ear). By the time that the scaffolding had dissolved, the cartilage had enough structural integrity to support itself.

Human Use?

The original cartilaginous structure that looked like a human ear was never transplanted onto a human being. This is because it was full of cow cells and would have been rejected by a person's immune system.

But in 1994, the same tissue technology had already been used on 12-year-old Sean G. McCormack, who had been born with the very rare Poland's Syndrome. He had absolutely no bone or cartilage on his left chest, his heart and lungs protected only by skin. You could look at the left side of his chest and actually see the skin move up and down as his heart expanded and contracted. This was a potential problem every day of his life, especially in his beloved sport of baseball. He was a star pitcher and a single ball to the chest could kill him. He needed a chest wall.

The Vacanti brothers started with their synthetic biodegradable polymer in a flat round disc roughly the size of a CD. This disc was moulded to the shape of his chest. They then seeded McCormack's own cartilage cells into this 'chest plate'. They implanted the seeded cartilage plate in his chest and it grew with him. Today he is healthy.

And like the mouse with the 'human' ear, there was absolutely no genetic engineering involved — only genuine scientific invention.

Non-biological Implants

Non-biological implants have been used in human beings for many years. They include metal/ceramic hip and knee joints and metal/silicone rubber valves in the heart.

The first problem is that they don't grow with the patient, which limits their use in children. The second problem is that they are not self-repairing, unlike many tissues in the body. As a result, many implanted joints have a life span of some 15 years.

In the USA, surgeons perform some 250 000 knee joint meniscus operations each year, because of routine injury and wear and tear. The vast majority of the operations simply clean up the mess and do not involve implants of metal. So the knee is left in a damaged state.

If only it were possible to transplant biological knees made from the patients' own cells.

Natural Bladder

In April 2006, *The Lancet* published a report about a 1999 procedure in which bladders were grown for seven patients (aged 4–19) with spina bifida. This was the first time that internal organs had been grown on the laboratory bench and then implanted in human beings.

It was a mammoth effort spread over a few decades. First, the team had to make the biodegradable hollow bladder scaffold. The next stage involved implanting two different types of cells into the scaffold — urothelial cells (which line the human bladder) and smooth muscle cells (which make the bladder contract). The team had worked for over a decade before they found the right ratios of different growth factor chemicals to make the two different types of cells grow.

Once they implanted the bladders, they had to use catheters to empty them. However, within a few months, the bladders had married themselves to nearby nerves and blood vessels and had begun to work normally. Finally, they wanted to wait seven years to make sure that they kept working normally.

Pots and Pans

Trisha Gura wrote a lovely description of today's unimaginative implant materials in the *New Scientist* of 29 May 1999. Her article is a very nice introduction to Tissue Engineering.

She noted that the vast majority of current implant materials used in the body 'bear no resemblance to the real thing, such as plastic, metal and porcelain, or the higher-tech Teflon, Dacron (called Terylene in Britain) or Goretex — in other words, the stuff of pots and pans, bras and expensive raincoats'.

Ear Seen Around the World

There were a few reasons for choosing the external ear as a first project to tissue engineer. Children are born without ears and ears get damaged, but there was another reason. The human ear is mostly a single type of cell and its product is cartilage. There are very few areas of the body that are easy to get to — and made of just one type of cell.

This technology, officially called Tissue Engineering, has been slow to take off. So far, the successes in Tissue Engineering have involved single cell types.

Cartilage doesn't need much oxygen, so the small numbers of blood vessels that the mouse grew into the synthetic biodegradable scaffold were sufficient. Solid organs that need huge numbers of new blood vessels are simply too difficult for today's Tissue Engineering technology. Scientists can grow blood vessels that look fine — on the laboratory bench. Unfortunately, these blood vessels simply explode when subjected to normal blood pressure. The liver has another difficulty over and above infiltrating it with blood vessels. It has about six different types of cells that have to be organised very precisely.

And note the famous 'ear on the back of the mouse' did not have a layer of skin on it. It was totally encapsulated by the skin of the mouse.

Growing cartilage (a single tissue) is difficult enough, but growing cartilage with a layer of skin on it is still too difficult.

Auriculosaurus

The external part of the human ear is called the auricle. Its job is to hold up your glasses and to collect and focus sound that comes from in front of you and out to the side, and deliver it to the ear canal. The ear drum is a few centimetres inside the ear canal.

Once the famous picture of the mouse made its way around the world, it was rapidly given the name Auriculosaurus.

Seaweed

A major breakthrough came by staring at seaweed.

A big problem in Tissue Engineering was (and still is) delivering the blood supply deep into growing organs. It's still difficult to grow blood vessels.

In 1986, Joseph Vacanti took his family on a beach holiday. He sat on a jetty, contemplating this problem. Suddenly, he noticed a thin sheet of seaweed floating in the water. It didn't need many pipes or blood vessels running through it delivering nutrients and removing wastes, because it was so thin. This meant that every part of the seaweed was close to the nourishing ocean. That's why his team chose to work with thin structures like ears and chest plates in the early days of Tissue Engineering.

References

Atala, Anthony, et al., 'Tissue-engineered autologous bladders for patients needing cystoplasty', *The Lancet*, 15 April 2006, pp 1241-1246.

Cao, Y.L. and Vacanti, J.P., et al., 'Transplantation of chondrocytes utilizing a polymer-cell construct to produce tissue-engineered cartilage in the shape of a human ear', *Plastic and Reconstructive Surgery*, 1997, Vol 100(2), pp 297-302.

D'Agnese, Joseph, 'Brothers with heart', *Discover*, July 2001.

Gura, Trisha, 'Custom-made for you', *New Scientist*, 29 May 1999.

Milloy, Steven, 'The Green's ear-ie ad groups use scare tactics to fight technology', *Washington Times*, 10 December 1999.

Pothula, V.B., et al., 'Otology in ancient India', *Journal of Laryngology. and Otology*, March 2001, Vol 115, pp 179-183.

MICROWAVES DAMAGE FOOD

A few years ago, our family visited Lake Mungo, a World Heritage Site in far southwestern New South Wales. On a guided tour the ranger showed us the remains of a barbecue that the local Aborigines had enjoyed about 23 000 years ago. It is still the oldest cooking site we have seen.

When it comes to cooking food, there hadn't been any real advances since the cook-up in Lake Mungo, until the arrival of the microwave oven 50 years ago. Microwaves and Mungo just follow different pathways to the same end result — applying heat to cook food. However, thanks to some very poor science, many people wrongly believe that cooking food in your kitchen 'nuker' will not only destroy the nutritional goodness in your food but will also mutate the previously benign nutrients into cancer-causing chemicals.

'Evils' of Microwaved Food

Lita Lee's book, *Health Effects of Microwave Radiation — Microwave Ovens*, makes these claims. She writes that microwaved food suffers 'decreased bioavailability of vitamin B complex, vitamin C, vitamin E, essential minerals and lipotropic factors in all foods tested'.

Many web sites have jumped on the bandwagon, making even wilder claims about microwaved food such as: '… causes long-term permanent brain damage by shorting out electrical impulses in the brain' and '… minerals, vitamins and nutrients of all microwaved food is (sic) reduced or altered so that the human body gets little or no benefit'. They use pseudoscientific gibberish such as '… a degeneration and destabilisation of the external energy activated potentials of food utilisation within the process of human metabolism'.

Indeed, the microwave oven is further demonised by the claim that it was invented by the Nazis and therefore must be bad. This is totally wrong. The microwave oven was accidentally invented by Dr Percy L. Spencer of the Raytheon Corporation in the USA in 1946.

The funny thing is that the Sun (which is surely 100% natural) emits microwaves. But this doesn't prove to be any inconvenience to the anti-microwave bandwagon. Instead, they make the outrageous claim that the Sun's microwaves are '… based on principles of pulsed direct current (DC) that don't create frictional heat; microwave ovens use alternating current (AC) creating frictional heat'. This particular quote is another example of pseudoscientific gibberish, using scientific phrases randomly strung together.

Bad Science

A fairly large (and inaccurate) body of literature claims that microwaved food is depleted in nutrients, and/or full of carcinogenic chemicals. Two studies from this literature are quoted continually. The first one appeared in *The Lancet* on 9 December 1989, while the second was a study by Hertel and Blanc in Switzerland.

The *Lancet* study, 'Aminoacid isomerisation and microwave exposure', is not a formal, peer-reviewed article, but merely a short letter to the editor (see *Peer Review* on page 143). The authors microwaved milk and found that a certain protein (L-proline) changed its shape to D-proline. This was a real worry because D-proline, in large concentrations, is toxic to the kidney and liver. As a result, health authorities around the world redid the same experiment.

The overall finding was that this was not relevant to the home heating of milk, because the scientists had exposed the milk to much greater levels of microwaves than are used in the home. For example, applying gentle radiant heat to thin slices of bread will give you delicious safe-to-eat, golden-brown toast. But burning slices of bread with lots of radiant heat will give you burnt toast. This blackened toast now contains chemicals that can give you cancer. Therefore, the degree of heating is important.

Really Bad Science

The second study — carried out by a retired chemist called Hans Hertel — looked at levels of various chemicals and cells in the blood of volunteers who ate microwaved and non-microwaved food over a two-month period. An article about the study, written by a René d'Ombresson, was published in a journal owned by a certain Franz Weber. It was modestly called *Journal Franz Weber*. This journal is definitely not peer reviewed. If it had been, four big problems with the study might have been noticed.

First, there were only eight volunteers. This is such a small number that it's impossible to get any statistically significant results. Second, for some of his measurements, Hertel used a bizarre method currently unknown to Western science, namely '… bacterial bioluminescence, which allows the degree of stimulation or inhibition of bacteria in the blood to be measured'. Third, when he used conventional tests, none of the blood analysis results actually fell out of the normal range of variation! Fourth, the volunteers, who all ate macrobiotic food, came to the study with a low-grade anaemia; they actually began the study being a little unwell.

On top of all this bad science, the study concluded that '… it is certain that you will die from cancer …'.

Yes, it is true that we will all die someday, but it's very unlikely that the cause in every case will be cancer induced by eating microwaved food.

Many millions of people have eaten microwaved food and later died from various conditions. Some of them died from cancer, but a much larger percentage died of conditions other than cancer. There is no link between eating microwaved food and dying of cancer.

WMD in the kitchen

Microwaving food retains vitamins and minerals better than other forms of heating.

Good Science

Because the microwave oven has now been used for cooking for more than 50 years we do have some scientific data on the subject.

We do know that water-soluble vitamins (e.g. C and the B group) are broken down by heat. They also dissolve in water. So some vitamins are lost by both microwaving and boiling. But when you do the measurements, it turns out that there are more vitamins left in microwaved vegetables than in boiled vegetables. And when you compare microwaved and steamed vegetables, you find that they have roughly similar amounts of these water-soluble vitamins.

With vitamins that are insoluble in water (e.g. A and D), microwaving has a clear advantage over boiling. The water-insoluble vitamins are exposed to less heat for a shorter time, so they survive better than those that are boiled.

Minerals (e.g. sodium and potassium) are mostly soluble in water, and so are retained better when microwaved.

The verdict? Microwave heating is better for retaining vitamins and minerals in food than other forms of heating.

Fats and carbohydrates are basically unaffected by both regular heating and microwave heating.

Proteins suffer less oxidation in a microwave oven (lower temperatures, shorter time) than in conventional cooking, and therefore the quality of protein is higher. Indeed, the lack of browning is proof that the heating is gentler.

Killer Barbecue

But there is one thing you can be sure of. The delicious slightly burnt crispyness that you get with barbecued meat definitely contains chemicals that can (in studies involving lots of rats) cause cancer. Even so, most of us ignore this scientific finding and eat barbequed meat with no regrets.

So heat, eat and be merry ...

Peer Review

The process called 'peer review' is not perfect, but it's not bad. It tries to ensure that what an author publishes in a scientific or medical journal is not rubbish.

The author's paper is sent to people who are experts in the same field (i.e. they are peers of the author) to decide whether it is good enough for publication. These peers then review the paper, checking for validity, errors or omissions, grammar, scholarly use of theory, reproducibility of results, deliberate faking, etc. The paper, if sufficiently authoritative, then becomes a major article in a journal, such as *Nature*, *Science*, *The New England Journal of Medicine*, *The Lancet*, etc. Such articles represent a significant advancement in our knowledge.

Major articles in scientific and medical journals have to be peer reviewed. 'Letters' do not, because they serve a different purpose from major articles. A letter may be just an observation. For example, if one doctor has noticed that some patients with a strange immunodeficiency often have a blueish rash on their legs, they may ask if any other doctors have noticed this.

Many so-called 'articles' quoted by advocates of junk science are actually just letters.

Lake Mungo

At Lake Mungo, human remains have been dated to 40 000 years ago, while stone tools have been dated to 50 000 years ago. Today the area is very arid. But back around 44–26 000 years ago the world was in an Ice Age, and Lake Mungo was prime real estate, the Land of Milk and Honey — mild climate, 6 m deep and brimming with fish.

In this fertile paradise of food and water, edible plants such as water lilies and spike-rushes flourished in the water. The nearby sand dunes and plains were an abundant and very diverse mix of woodlands, scrub and grasslands. The Aborigines would have eaten Quandongs, acacia seeds, Native Bush Tomatoes and the succulent fruits of the native cherry and the Ruby Saltbush. We know that they ate Golden and Macquarie Perch, mussels and Murray Cod from the waters. The land animals that they ate included the large Short-faced Kangaroo, the cattle-sized Zygomaturus, and the great flightless bird, Genyornis — a bird much bigger than an Emu, with legs like those on a horse. The Aborigines also ate smaller marsupials such as the Thylacine (Tasmanian Tiger), Hairy-nosed Wombats, the ancestors of both the Grey and Red Kangaroos, and smaller animals like Bilbies and Bettongs.

The ranger, Tony Woodhouse, took us for a little stroll and asked us what we thought of a little dark smudge on the cracked, grey claypan. We weren't very impressed, and said that it looked like a bit of dirt that had blown in. We were really amazed to find that we were looking at a 23 000-year-old BBQ — possibly one of the oldest known barbies on the planet! The cracking of the clay was partially due to the heat of the fire. As we looked more closely, we could see that the blackness was from the burnt coals of this ancient barbie.

And then he pointed out little white bones, roughly the size of a fingernail, which I had previously ignored. These bones were otoliths — bones used in the balance centres of fish. They normally rest in a hollow cavity lined with sensitive hairs that all point in towards the centre. The otolith rests on only a few of these hairs. The hairs send electrical signals to the fish's brain, helping it work out which way is up. There are two special things about otoliths — they are unique to each type of fish and they get bigger as the fish grows. So Tony was able to tell us that 23 000 years ago, some Aborigines barbecued a certain type of fish about one metre long.

I was so impressed that I had to sit there, trying to soak up the feel of probably the oldest barbie on the planet.

Of course, over this immense period of time of some 50 000 years, there were a few small periods of droughts. We know this from the layers of sand and dirt left behind.

But the good times in this Land of Milk and Honey had largely come to an end about 18 000–20 000 years ago. At that time, the Little Ice Age had a final hissy fit, and the climate changed to dry and cold — about 6–10°C colder than today. The lakes dried out and the landscape became arid again.

References

Bowler, James M., et al, 'New ages for human occupation and climatic change at Lake Mungo, Australia', *Nature*, 20 February 2003, pp 837-840.

d'Ombresson, René, 'Microwave ovens: a health hazard. Irrefutable proof', *Journal Franz Weber*, 19th Issue, Jan/Feb/March 1992, pp 3-10.

Lubec, G., et al, 'Aminoacid isomerisation and microwave exposure', *The Lancet*, 9 December 1989, pp 1392, 1393.

Stanton, Dr. Rosemary, 'The fear factor: lack of understanding has given microwave cooking a bad name', *Australian Doctor*, 15 July 2005, p 37.

IDENTICAL SNOWFLAKES

We don't see a lot of snow in Australia because it tends to snow at the higher latitudes closer to the Poles (above about 35°). Most of Australia is closer to the equator than this. Although we don't have a lot of experience with snow, like most of the rest of the world we do think that 'no two snow crystals are the same' — and, you guessed it, this is yet another myth.

Snow can happen at low latitudes, if the mountain is high enough. An example is Mt Kilimanjaro, which is only a few degrees from the equator. (However, it seems that the equatorial ice may soon vanish, thanks to Global Warming.) Mt Ranier near Seattle holds the record for the heaviest snowfall in a season — 1000 inches (roughly 25 m) in 1971–1972 — while Colorado had the highest single-day snowfall of 76 inches (193 cm) in 1921.

Snow Making 101

Snow forms when water vapour is exposed to very low temperatures (such as at the tops of clouds) and turns into ice.

The 'classic' snowflake has six spokes radiating out from a common centre, in the same way that the spokes of a bicycle

No two are the same, eh?

The 'Snowflake Man'
Wilson A. Bentley

In Wilson's snowflake recording career (some 40 years) he never found two snowflakes alike. That was until the talented scientist Nancy Knight came along and found two **identical** snowflakes in a Wisconsin snowstorm.

wheel radiate out from the central hub. However, this star pattern is only one of many possible shapes, which vary depending on temperature, closeness to other snowflakes, wind and so on. For example, if you steadily drop the temperature from 0°C to –25°C and keep everything else constant, you will see snowflake shapes running through hexagonal plates, needles, hollow prisms, plates, stellar dendrites (the 'classical' star shape), back to plates again, and finishing up with solid prisms.

The forces that make water molecules leap from one ice-crystal face to another seem to rely heavily on the local temperature. This helps explain why there are so many different shapes.

As a snowflake falls, it tumbles through many different environments. The snowflake that you see on the ground is deeply affected by the different temperatures, humidities, velocities, turbulences, etc. that it has experienced. Because most snowflakes would have slightly different flight paths, they would have different shapes. Of course, snowflakes that land near each other tend to have similar histories.

The Snowflake Man

Wilson A. Bentley, a farmer who was born, lived and died in the small town of Jericho in Vermont, was called 'The Snowflake Man'. He supported the 'all snowflakes are different' theory. Around 1884, at the age of 19, he became the first person to photograph a single ice crystal, by cleverly marrying a microscope to a camera using an adjustable bellows mechanism. In 1920, the American Meteorological Society elected him the status of Fellow. They also awarded him their very first research grant — $25 in recognition of his '40 years of extremely patient work'. He continued working in this field until his death in 1931, by which time he had taken 5381 'photomicrographs' of individual snowflakes. Towards the end of his life, he said that he had 'never seen two snowflakes alike'. From this arose the story that all snowflakes are different.

In 1988, the scientist Nancy Knight (at the National Center for Atmospheric Research in Boulder, Colorado) was studying cirrus clouds. During a snowstorm in Wisconsin her research plane collected snowflakes on a chilled glass slide coated with a sticky oil. Two of the snowflakes collected were identical (under a microscope, at least).

The snowflakes were hollow hexagonal prisms, rather than the classical six-spoked star shapes — but as far as snowologists are concerned, they were snowflakes. To be pedantic, they probably weren't identical if you were to look at the actual molecules. But at this level is anything identical?

Since the Earth was formed, about 4.5 billion years ago, around a million trillion trillion snowflakes have fallen — but Mr Bentley made his pronouncement after looking at only 5381 of them..

Reference

Edwards, Owen, 'Freeze Frame', *Smithsonian*, January 2005, pp 30, 31.

WALK THE PLANK

Pirates have been causing trouble for over three thousand years. They're famous for plundering, 'me hearty-ing' in broad nautical accents, wearing eye patches and making their enemies walk the plank. Plundering pays the bills, an accent makes you easily identifiable and a patch is handy for keeping dirt out of an empty eye socket. But did people really 'walk the plank' in the good old days?

Despite all the plank-walkers you may have seen in pirate movies, including Keira Knightley in *Pirates of the Carribean*, walking the plank is a relatively modern invention — and hardly ever happened.

History of Piracy

The *Encyclopaedia Britannica* defines 'piracy' as a 'robbery or other violent action, for private ends, and without authorisation by public authority, committed on the seas or in the air'.

The first mention of piracy is on a clay tablet dating back to the Pharaoh Akhenaten in 1350 BC. It describes piracy in the Mediterranean Sea. Later, Greek merchants mention piracy as part of the cost of being a maritime merchant. The Roman historian Polybius was the first to use the word 'pirate' (*peirato*), around 140 BC.

In 75 BC, the future emperor, Julius Caesar, was captured by pirates and held to ransom for 38 days. He behaved more like a guest than a prisoner while held in the Greek islands. He wrote poetry, rebuked his captors when they didn't pay enough attention to his poems and ordered them to stop partying when he wanted to sleep. After his ransom was paid, he set off for reinforcements and returned for revenge. He arrested his former pirate captors, bringing them back to the mainland to be hanged.

Around 100 AD, the Greek historian Plutarch defined pirates as those who attacked ships or cities without legal authority. By this time, there were so many pirates (on over 1000 ships) that they almost stopped maritime trade in the Mediterranean. They had captured or raided some 400 cities. The Roman commander Pompey was given authority to do whatever was necessary to get rid of the pirates. Using Roman military organisation and power, he did so in just three months.

Privateer — Profit and Patriotism

One special type of pirate was the privateer who combined profit and patriotism.

Privateers were 'legal' pirates commissioned by one country to attack the ships of another. The advantage to the commissioning country was that they didn't have to pay for the expensive upkeep of a large permanent navy — they could 'hire' privateers as needed. The privateers got a percentage of the take to keep them happy.

The privateer's protection was the Letter of Marque, issued by the government to show that the privateers had been legally authorised to attack enemy shipping. This supposedly stopped the enemy from trying privateers as pirates if they were caught. However, in keeping with their spirit of free enterprise, some privateers had a bet each way by holding a Letter of Marque from each of the two opposing sides.

Surprisingly, England did not have a large naval fleet for many years, making do with privateers. Privateering was probably initiated by King Henry III in 1243, when he engaged three ships to attack French shipping.

'This be t' plank i would like you t' walk on'

Pirates were generally a pretty ugly bunch. They really didn't set about wasting time having folk 'walk the plank' ... instead they usually just killed them quickly ... and in cold blood.

Privateering had a long and renowned history. Famous privateers included Sir Francis Drake (who defeated the Spanish Armada in 1588), William Dampier (who named Shark Bay in Western Australia in 1699) and Jean Lafitte (who defended New Orleans against the British in 1812).

Captain Blackbeard (Edward Teach, or Thatch) began his bloody career as a British privateer. After the war with the French and the Spanish ended he simply continued attacking ships, by taking over the *Queen Anne's Revenge*. Blackbeard set up his base in North Carolina (at the time a British colony). From here he could easily attack ships travelling along the American coast. He was caught and executed in 1718.

Spain was the last country to hire privateers, in 1906.

Golden Age of Piracy

The supposed Golden Age of Piracy occurred between 1680 and 1730. This period stands out because of a combination of factors.

First, privateering was quite a legitimate line of work at the time. As the various wars gradually wound down, the captains simply continued privately in their line of business with their ships and well-trained crews.

Second, the Caribbean was filled with Spanish ships shuttling back to the mother country, loaded with all kinds of riches, including gold and silver. These ships were attractive targets.

Third, John Esqemeling wrote an extremely popular book about pirates called *The Bucaneers of America*, which combined a mix of romanticism and fact. It was published in Dutch in 1678, in Spanish in 1681 and, finally, in English in 1684.

Fourth, many of the pirate ships ran on an early form of democracy. The captain was often elected by the crew, and the entire crew shared in the profits. This was a revolutionary, and very attractive, concept for a time when people accepted the Divine Right of kings to do anything they wanted.

Modern Piracy

Piracy still happens at sea today, mainly in the South China Sea and off the African coast. Modern sea pirates tend to use small, fast motorboats and small, powerful arms such as mortars, machine guns and rocket-propelled grenades. In 1996, pirates boarded the tanker *Succi*, forcing the crew into a lifeboat. The *Succi* then vanished.

The good news, according to the International Maritime Bureau, is that pirate attacks against international shipping dropped by 27% in 2004 (325 incidents in 2004 vs 445 in 2003).

The bad news is that the Strait of Malacca — home to half the world's piracy — is still very vulnerable. Running between Malaysia and Sumatra, it is the shortest sea route between the Indian and Pacific Oceans. It is about 1000 km long, but less than 2 km wide at its narrowest choke point. Each year, 50 000 ships pass through the strait, carrying two-thirds of the world's liquefied natural gas, half of the world's crude oil and one-third of its commerce. These merchant ships are the classic 'low-hanging fruit'. They are slow, unarmed and loaded with valuable goods — anything from cars and ore to oil and Japanese nuclear waste bound for recycling in Europe.

Walking the Plank — Zero to Five

But what of 'walking the plank'? Most pirates just killed their enemies in cold blood as quickly and unromantically as possible.

Bartholomew Roberts (1682–1722), also known as Black Bart, was reputed to have made (only) one enemy walk the plank. Probably the most successful pirate ever, Black Bart captured 400 ships and stole £50 million in just two and a half years. He did not allow gambling on his ships, but was violent to the point of psychosis and tortured his victims. He was killed on 10 February 1722.

Walking the plank was first mentioned in 1837, in Charles Ellms's *The Pirates Own Book*, an American story book for boys. Ellms claimed that the American amateur pirate, Stede Bonnet (c. 1688–1718), made his enemies walk the plank. It is also described in Chapters 1, 28 and 31 in Robert Louis Stevenson's book, *Treasure Island*, written in 1881.

Despite popular perception, the historical records document only a few genuine cases of walking the plank — somewhere between zero and five.

The phrase 'walking the plank' conjures up such wonderfully graphic images — especially to anyone familiar with Peter Pan and Captain Hook. It now has a firm place in the English language, e.g. 'the CEO had to walk the plank'.

As the historian Michel-Rolph Trouillot said, 'it's not just historians who write history, but popular authors, television and movies as well …'

Articles of Black Bart

The movie *Pirates of the Caribbean* refers to a single 'code' that all pirates followed. There was no such code, but there were many 'Articles'. Each ship had its own set of Articles, which everyone signed. This early form of Industrial Relations Workplace Agreement bound the crew and captain of a pirate ship. The following will give you some idea of how the Articles worked on Black Bart's ships:

- If the robbery was only betwixt one another, they contented themselves with slitting the ears and nose of him that was guilty, and set him on shore, not in an uninhabited place, but somewhere where he was sure to encounter hardships.
- No person to game at cards or dice for money.
- The lights and candles to be put out at eight o'clock at night: if any of the crew, after that hour, still remained inclined for drinking, they were to do it on the open deck.
- No boy or woman to be allowed amongst them. If any man were to be found seducing any of the latter sex, and carried her to sea, disguised, he was to suffer death.
- To desert their ship or quarters in battle, was punished with death or marooning.
- No striking one another on board, but every man's quarrels to be ended on shore, at sword and pistol.
- No man to talk of breaking up their way of living, till each had shared £1000. If, in order to do this, any man should lose a limb, or become a cripple in their service, he was to have £800, out of the public stock, and for lesser hurts, proportionately.
- The captain and quartermaster to receive two shares of the prize: the master, boatswain, and gunner, one share and a half, and other officers one and a quarter.

Gallows

The gallows was the 'traditional' Admiralty punishment for the captured pirate. He was to be hanged on the shore, below the high tide mark. The gallows were supposed to be left there after the hanging until three high tides had washed over them.

References

Burnett, John S., 'The next 9/11 could happen at sea', *The New York Times*, 22 February 2004.

Encyclopaedia Britannica, Ultimate Reference Suite DVD, 2006 — 'piracy'.

Higgins, Glynn, 'Walking the plank', *Fortean Times*, October 2004, p 71.

Luft, Gal and Korin, Anne, 'The modern pirates', *The Weekend Australian Financial Review*, 23-28 December 2004, pp 5, 10.

A FOUR-LEAF CLOVER – PLUCKIN' LUCKY

Whenever my children go on a picnic, they go looking for a four-leaf clover because they have come to believe that finding one will bring them luck. Each of the four leaves has its own special type of luck — fame, wealth, a faithful lover and glorious health. An essential part of believing this myth is that four-leaf clovers are very rare. In reality, finding a four-leaf clover is not very difficult — it just depends on where you are.

By the way, one popular song has a different interpretation of the four leaves. It runs:

'I'm looking over a four-leaf clover
That I overlooked before.
One leaf is sunshine,
The second is rain,
The third is the roses that bloom in the lane,
No need explaining the one remaining,
It's somebody I adore.'

In another interpretation the leaves stand for faith, hope, love and luck. There are many interpretations, but all of them involve some kind of good fortune.

'I'm looking over a four-leaf clover'

It is commonly believed that finding a four-leaf clover will bring luck in the form of fame, wealth, a faithful lover and good health. BUT ... it could simply mean the plant is stressed!

Clover 101

There are about 300 different species of clover in the genus *Trifolium*. Cultivated species of clover originated in Europe, but have successfully spread to temperate regions around the world.

Historically, farmers have grown lots of clover because it is high in calcium, phosphorus and protein and livestock love to eat it, either fresh or dried as hay. Clover is also popular with farmers because it slows down erosion by holding the soil together, can dump up to 170 kg of nitrogen per year in each hectare of land, and can make other nutrients more available to the next lot of crops grown. So, clover improves the soil.

The more significant agricultural species are red clover *Trifolium pratense*, alsike clover *T. hybridum* and white clover *T. repens*. Red clover produces an oval-shaped, purple flower 2–3 cm across, while alsike clover (also called Alsatian or Swedish clover) has spherical, rosy-pink flowers. White clover, a popular seed in lawn-grass mixtures, has a white flower that often has a hint of pink.

All in all, clover is a highly desirable crop. But why is it supposed to be lucky?

Lucky Clover?

We don't really know how the 'lucky four-leaf clover' myth came into being.

One explanation dates this myth back to the biblical story of the expulsion of Adam and Eve from Paradise. Apparently Eve took a four-leaf clover with her to remind her of luckier times. So having some clover growing in your garden will bring you luck.

Another explanation comes from various Irish legends. The Druids — from the time of the ancient Celts of Gaul, Britain and Ireland — used to assemble in their sacred oak forests several times each year to settle disputes, and have celebrations. One of their last rituals before returning home was to search for mistletoe and the rare four-leaf clover. They believed that mistletoe would bring peace to their household, and that whoever held a four-leaf clover could see otherwise hidden demons. Using the power of the lucky clover, they could then stop the demons by invoking the appropriate spells.

Rare No More

But four-leaf clovers are not really rare.

Firstly, five American species (*Trifolium andersonii, T. lemmonii, T. thompsonii, T. gymnocarpon* and *T. macrocephalum*) and two European species (*T. lupinaster* and *T. polyphyllum*) occasionally have a natural tendency to produce individual mutant clover plants with four or nine leaves.

Secondly, the rarity of the four-leaf clover was destroyed in the 1950s, when horticulturists bred a species of clover that would always sprout four leaves.

Thirdly, to even further erode the value of the four-leaf clover, any of the 'regular' species of clover that usually have only three leaves can generate more than the three 'standard' leaves if they are stressed. Stresses include environmental factors such as herbicides, diseases, viruses, aphids, dry conditions and even cold nights or warm days. Australian climates are usually different from the climates in which *Trifolium* originally evolved. These climatic differences can be enough of a stress to force some clover plants to pop out four leaves. So your location can help turn a regular three-leaf clover into a four-leaf clover.

And if all else fails, you can pick any three-leaf clover and then surreptitiously slit one of the leaves into two smaller leaves to make your own special four-leaf clover.

How To Be Lucky

Professor Richard Wiseman holds Britain's only chair in Public Understanding of Psychology, at the University of Hertfordshire. He has written a book, *The Luck Factor*, in which he presents four simple principles that will help you change your luck.

First, maximise your lucky opportunities. People who are consistently lucky will create, notice and act upon random opportunities.

Second, follow your hunches and listen to your gut feelings and intuition. So take a mental step backward and clear your mind of distracting rubbish to boost your natural intuitive abilities.

Third, expect good fortune. This faith can help you work through the patches of bad times that inevitably come from time to time. It can also help you interact with other people in a positive way.

Finally, when bad luck comes (as it occasionally must), turn it into good luck. For example, if you break your leg, remind yourself that you could have broken something worse. Then, stop dwelling on the bad luck and take control of the situation.

Through his various studies over the years, Professor Wiseman has found that unlucky people have almost no insight into why they are unlucky.

In one study, he asked both lucky and unlucky people to count the number of photos in a newspaper that they were given. The unlucky people took two minutes, the lucky people took seconds. Why? The second page of the newspaper had a message, written in type about 5 cm high, and filling over half the page. The message said, 'Stop counting. There are 43 photographs in this newspaper'. The unlucky people tended to miss this huge message, while the lucky people tended to see it.

References

Brasch, Dr. R., *How Did It Begin*, Melbourne: Longmans, Green & Co. Ltd, 1965, pp 4, 5.

Panati, Charles, *The Extraordinary Origins of Everyday Things*, New York: Harper & Rowe Publishers Inc, 1987, p 8.

'Pluck for luck', The Last Word, *New Scientist*, 14 September 2002, p 73.

Wiseman, Richard, 'Luck's a fortune — and easy to learn', *The Sydney Morning Herald*, 17 January 2003, p 1.

BOILING FROGS

If you have listened to enough motivational speakers, politicians, management consultants, environmentalists or religious clergy speak, you will almost certainly have heard the story of the frog in hot water. It claims that if you put a frog in boiling water it will immediately jump out. But if you put the frog in water at room temperature, and then gradually heat the water to boiling, the frog doesn't realise what's happening and gets cooked to death.

This neat (but gruesome) image is often used to warn you about microscopic changes — usually imposed by the Forces of Evil — that will gradually and relentlessly whittle away your rights, your freedom, your money or your distance from the Devil. Because each change is so small, you supposedly don't notice it happening.

Everybody Uses It

For example, members of the American Extreme Right use the frog-in-the-water image to accuse the Bush Administration of stealing their money (through federal income tax), their privacy and their rights. The American Left invoke the same image to illustrate their ever-growing financial dependence on Washington, and the ever increasing use of surveillance technology to erode their privacy.

Economists use this story in Economics 101 lectures, claiming that the average citizen doesn't notice day-to-day economic changes and is oblivious to any major economic shifts.

Environmentalists do the same to show that the environment is continually being eroded bit by bit, with major damage going unnoticed until it is too late. A good example is the classic London pea-soup fog. In 1952, one such fog killed 4000 Londoners. Gwynne Dyer wrote on http://DAWN.com: 'Drop a frog into a pot of boiling water, and he will promptly hop out again. Put him into cold water, bring it to the boil slowly, and he'll sit there until he dies. Londoners had grown used to the fogs, and didn't realise they were killing people because the victims weren't clutching their throats and falling over in the streets.'

Dr Julie Jonassen, Assistant Professor of Physiology at the University of Massachusetts Medical Center in Worcester, uses the frog story to discuss domestic violence. 'If you take one bucket and bring the water to a boil and then throw the frog in it will immediately die. If you take the second frog and put it in the second bucket and slowly turn up the heat it will adjust to the temperature until it dies. This is also true in abusive relationships where the woman will leave early on if she is hit but in a longer relationship she will stay,' she said.

Even More Use It

Carter McNamara uses the story to explain the ninth of his 10 Myths of Business Ethics: *Our organisation is not in trouble with the law, so we're ethical*. He rebuts this myth with: 'One can often be unethical, yet operate within the limits of the law, e.g. withhold information from superiors, fudge on budgets, constantly complain about others, etc. However, breaking the law often starts with unethical behaviour that has gone unnoticed. The "boil the frog" phenomenon is a useful parable here: If you put a frog in hot water, it immediately jumps out. If you put a frog in cool water and slowly heat up the water, you can eventually boil the frog.'

Corporate consultants cite the frog fable as proof that companies change so slowly that most of the employees, and often the managers, don't notice the change and cannot deal with it. And other consultants worry about the slow invasion of government into business. John Semmens writes, 'Instead, like the frog that never

Damn I'm HOT!

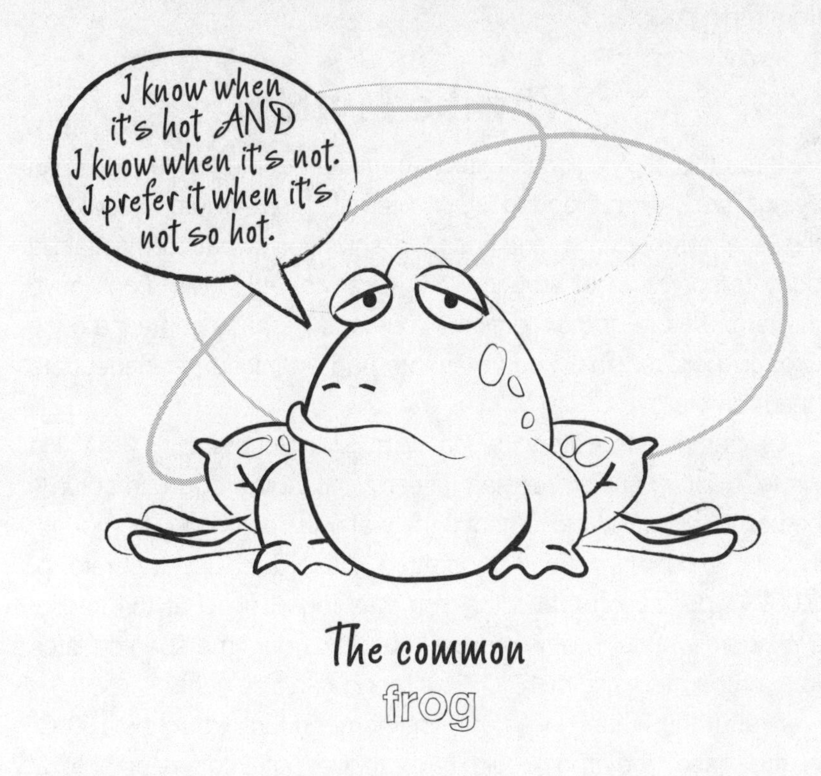

jumps from water brought slowly to a boil, business management has been unable to decide at what point the (US Government) destruction of managerial options will be fatal to the business enterprise'.

And concerned parents agonising over R-rated movies say: 'Our youth are like the frog in the boiling water ... they are being acclimated to the 'hot' water and don't even realise they will soon boil to death.'

Even real estate agent Mollie Wasserman of Boston says, 'If you throw the frog into boiling water, its nervous system will tell it to jump out. But if you put it in a pan of tepid water and turn the heat up, the frog will boil away. And if agents don't get smart, they will boil away.'

Political commentator Edip Yuksel uses it to describe the gradual acceptance of corruption. He says, 'People may ... suffer from "Boiled Frog Syndrome" ... Gradual corruption in political or social bodies can have fatal consequences without eliciting reaction from most people.'

Why the Myth?

Perhaps this story began because frogs are 'cold-blooded'. Human beings are 'warm-blooded'. Our internal 'thermometer' measures the local temperature, and then we shiver or sweat to maintain a body temperature of around 37°C. But 'cold-blooded' frogs have the temperature of their local environment. Perhaps someone once wrongly believed that frogs therefore had an inferior or inadequate 'thermometer'.

Or perhaps the story began with E.M. Scripture in 1897. He wrote, quoting earlier German research, '... a live frog can actually be boiled without a movement if the water is heated slowly enough; in one experiment the temperature was raised at the rate of 0.002°C per second, and the frog was found dead at the end of $2^1/_2$ hours without having moved.' The $2^1/_2$ hour time span equates to a temperature rise of 18°C. This report can't be right.

First, if the water boiled, the final temperature would be 100°C. In this case, the frog would have to be put into water already heated to 82°C (100 – 18 = 82). Surely, the frog would have died immediately in water as hot as 82°C. Second, Scripture wrote that the frog had not moved. How do you convince a frog not to move for $2^1/_2$ hours?

Why it's a Myth

There are two big problems with the frog-in-water story.

First, a frog can't jump out of boiling water.

Do you remember what happened the last time you dropped some egg white into boiling water? The proteins coagulated into a mess of thin white strands. Unfortunately, the proteins in the frog's skinny legs would do the same thing. The frog wouldn't be able to jump anywhere. In fact, it would die from its injuries. Dr George R. Zug (Curator of Reptiles and Amphibians at the National Museum of

Natural History in New York) and Professor Doug Melton (of the Harvard University Biology Department) both agree on this point. It is not a question of the frog not wanting to jump, but not being able to.

Second, a frog would notice the water getting hot.

Dr Victor Hutchison, a herpetologist and Emeritus Research Professor of the Department of Zoology at the University of Oklahoma, has dealt with frogs all his professional life. Indeed, one of his current research interests is: 'the physiological ecology of thermal relations of amphibians and reptiles to include determinations of the factors which influence lethal temperatures, "critical thermal maxima" and minima, thermal selection, and thermoregulatory behaviour'. 'Critical thermal maxima' means the maximum temperature that the animal can bear.

Professor Hutchison says: 'The legend is entirely incorrect! The "critical thermal maxima" of many species of frogs have been determined by several investigators. In this procedure, the water in which a frog is submerged is heated gradually at about 2°F per minute. As the temperature of the water is gradually increased, the frog will eventually become more and more active in attempts to escape the heated water. If the container size and opening allow the frog to jump out, it will do so.'

So experiments show that the frog story is wrong. But in all my reading, I couldn't find any studies where the experimenters tried to slowly boil consultants, motivational speakers or politicians.

Squeaky Croaks

Fish and homing pigeons can 'see' electromagnetic fields, ants can 'see' polarised light, insects and rodents can 'smell' pheromones, so why can't frogs 'talk' using ultrasound?

One species of frog can: the Concave-eared Torrent Frog that lives in the Huangshan Hot Springs west of Shanghai. The frog does have concave ears, and there is a continuous torrent of water and sound in its mountainous environment. The scientist who discovered this strange ability said, 'Nature has a way of evolving mechanisms to facilitate communication in very adverse situations. One of the ways is to shift the frequencies beyond the spectrum of the background noise. Mammals such as bats, whales and dolphins do this, and use ultrasound for their sonar system and communication. Frogs were never taken into consideration for being able to do this.'

The ears are concave to better hear ultrasound. It was known that these frogs made high-pitched sounds like twittering birds. But it was Albert Feng from the University of Illinois who discovered that some of the frogs' sounds were way up high in the ultrasonic — over 128 kHz — or more than six times higher than a human being can hear.

References

Feng, Albert S., et al., 'Ultrasonic communication in frogs', *Nature*, 16 March 2006, pp 333-335.

Gibbons, Whit, 'Legend of the boiling frog is just a legend, but does have environmental value', *Athens Banner-Herald*, 12 December 2002.

http://DAWN.com, Gwynne Dyer, 16 December 2002.

http://www.libertyhaven.com/theoreticalorphilosophicalissues/protectionismpopulismandinterventionism/govbusiness.html — John Semmens, 'Government Regulation and Business Management'.

http://www.snopes.com/critters/wild/frogboil.htm — 'Boiled Beef'.

Scripture, E.W., *The New Psychology*, London: Walter Scott, Ltd, 1987, pp 300, 301.

Yuksel, Edip, 'Lottery elections: disinfecting democracy from lobbies', 1998, http://www.Yuksel.org/e/law/lottery/htm.

SEAT BELT TRAP

Some people will argue about anything. They will even tell you that seat belts don't really work, and that you're better off being thrown out of the car to safety.

If you really push them in the argument, they will tell you a heart-rending story that they know personally to be true. The story usually has a large number of 'innocents' (Girl Guides, nuns or bridesmaids) dying in a flaming or drowning wreck of a car — simply because they couldn't unclip their seat belts.

History of Seat Belts

Seat belts first appeared in the 1930s, when some US doctors installed lap belts in their own cars. In 1956, Volvo offered a diagonal chest belt as an accessory. In December 1970, Victoria became one of the first places in the world to make the use of lap-sash belts compulsory in both front and rear seats. All the other Australian states and territories had followed by the end of 1971.

Seat Belt Physics 101

In a collision, your car can stop so rapidly that it experiences a sudden deceleration of (say) 30 G. In plain English, everything becomes effectively 30 times heavier.

Click Clack, Front and Back

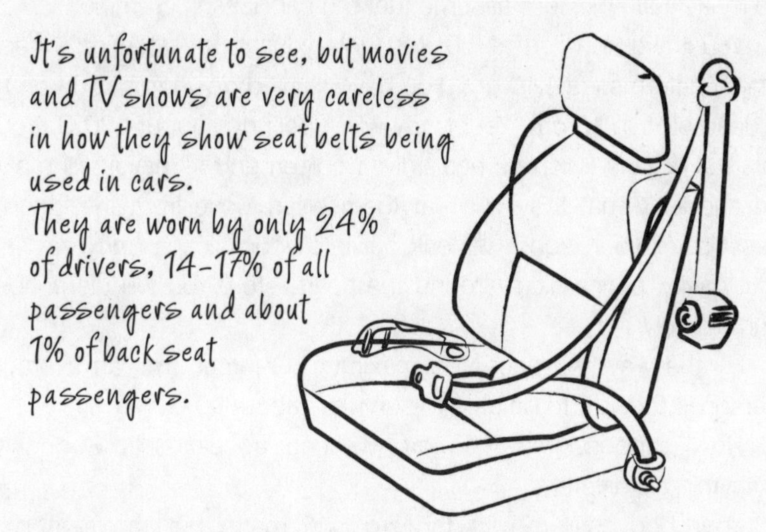

Simply put; wearing a seat belt means fewer deaths and injuries ... a certified 'no-brainer'!

Suppose that you are a passenger in the front seat. You are wearing a seat belt and holding a 10 kg infant on your lap. In a 30-G collision, the infant suddenly weighs 300 kg (the weight of four or five washing machines). You couldn't hang on to this many washing machines in a steady lift, let alone in a sudden jerk. As a result, the infant goes flying out of your arms and smashes into the windscreen.

What happens if you are not wearing a seat belt? In a 30-G collision, the average 70 kg person suddenly weighs about 2100 kg. There is no way that you can save yourself by stretching out your arms in a 30-G accident — your arms are simply not strong enough.

Seat Belt Physics 102

Suppose your car is travelling at 60 kph and comes to a sudden stop in zero distance. (Perhaps you ran into a fully loaded cement truck in an intersection, or you drifted over the centre line and ran head-on into a car of identical weight travelling at the same speed.) To make it simple, let's assume that the car doesn't crumple.

You travel forward at 60 kph until you hit the steering wheel, which deforms (say) 10 cm. You have slowed down from 60 kph to a dead stop in 10 cm. Your body will experience about 130 G. Your internal organs (such as heart, liver, spleen and kidneys) will move so suddenly that they will tear themselves loose from their blood vessels. All your blood will leak out into your various body cavities (gut cavity, pericardium around the heart, etc.). You will die almost immediately.

By the way, the accepted maximum G-forces that an average person is thought to be able to survive is about 40 G.

Now suppose that you are wearing a seat belt. Suddenly, everything is different.

Your body still moves forward one metre but the seat belt stretches as you do so. As you move forward, you slow down (from 60 kph to 55 kph to 50 kph, etc.) over 100 cm, not 10 cm. Ideally, you should have slowed down so much that you hit the steering wheel with almost zero velocity. Stopping over a 10-times greater distance reduces your deceleration to 13 G — so you live. You might have a nasty seat belt bruise on your chest, but you're alive.

Seat Belts — Lives and Money

Wearing seat belts in collisions means fewer deaths and injuries, and lowered costs to society. According to the Automotive Coalition for Traffic Safety in the USA, seat belts in cars have saved over 180 000 lives since 1975. Each year, seat belt usage saves the USA some US$50 billion in medical expenses, lost productivity and other costs related to injuries. On the other hand, the non-use of seat belts costs the USA about US$26 billion annually in terms of medical care, lost productivity and higher costs of taxes and insurance.

Each year in the USA about 45 000 lives are lost in road events. Wearing seat belts would cut this by about 10 000 lives. The American President could save 10 000 American lives each year by making seat belts compulsory in all American states.

Worldwide, seat belts reduce the chance of injury or death in a vehicle collision by up to 80%, depending on the type of collision (near-side impact, rollover, head-on, etc.).

In New South Wales, only 4% of drivers or passengers don't wear a seat belt. However, those 4% of lawbreakers account for 22% of car occupant deaths. In the Northern Territory in 2002, over 47% of car occupants who died were not wearing seat belts. Western Australian statistics show that you are 10 times more likely to be killed in a road crash if you don't wear a seat belt. Across Australia, in 33% of fatal accidents, the dead person was not wearing a seat belt and 20% of seriously injured occupants were also not wearing seat belts.

And look out for the passenger who does not wear a seat belt. In a collision, they bounce around inside the car. So they double the injury and death rate for the passengers who are wearing seat belts when they slam into them at high speed during a collision.

The Milwaukee Experience

Dr Shane Allen and his team at the Injury Research Center of Wisconsin in Milwaukee looked at seat belt data. Their results are typical of the studies in this field. They looked at all of the 23 920 crash victims who presented to hospital-based emergency departments in Milwaukee in 2002.

When non–seat belt wearers were compared to seat belt wearers, the trends were very clear.

The non–seat belt wearers were more likely to be male (56% vs 40%), to have used alcohol (17% vs 4%), to be involved in a single-vehicle accident (44% vs 22%), and to be younger (average age 31 years vs 38 years for seat belt wearers).

They were twice as likely to be admitted to hospital and up to four times more likely to suffer moderate to severe injuries to their whole body.

They were also three times more likely to die then those who did wear seat belts.

The Bottom Line

The lesson is absolutely clear and simple – driving can be dangerous. And you should always wear a seat belt. In the USA, there are six million car collisions each year, involving over 10 million people and causing about 45 000 deaths and 2.5 million injuries.

Practically every time, seat belts help you stay conscious during a collision. This means that you are more likely to be able to get out of the car after the collision, which is a great advantage if the car is in the water or on fire. Not wearing a seat belt during a collision means that you'll probably be knocked unconscious.

On the other hand, if you get thrown out of the car, you will almost certainly land on something hard and unyielding. Of course, the movies usually show the hero landing on something soft and cushiony, like a convenient haystack nearby.

Most people like to wear seat belts. Perhaps the people who don't wear seat belts don't like to be reminded of how dangerous driving really is. Even so, they should just belt up.

Slogans

Quite a few catchy advertising slogans have been used around the world to promote the wearing of seat belts. Here's a selection:

Click it or Ticket
Clickety Clack, Front and Back
Wear it, or wear $105
Wear it, or wear a fine
Driving unbuckled kills everyday people, every day
No seat belt, no chance
Belt up in the back

Movies and TV

Movies and TV shows are very careless in how they show seat belt use in cars.

Seat belts are worn by only 24% of drivers, 14–17% of all passengers and about 1% of back-seat passengers.

Strange Laws

The USA was slow to adopt seat belt laws. New Jersey was the second state to pass a seat belt law, which had a very unusual provision. Unfortunately, 40 other US states adopted this same provision. It came to be called Secondary Enforcement.

Secondary Enforcement means that a police officer cannot book you if the only law you have broken is not wearing a seat belt. But if they see you breaking a different traffic law, then they can book you for breaking the traffic law *and* for not wearing a seat belt.

A minority of states have Primary Enforcement seat belt laws, where the police officer can book you for not wearing a seat belt. But the majority of US states have Secondary Enforcement.

Alien Present

Suppose aliens came to Earth and offered us an amazing new personal transport technology. The annual cost would be US$518 billion, 1.2 million human lives and 50 million people injured. We would say, 'No, the price is too high, because human life is sacred.'

But that's what road crashes cost us worldwide in 2004 ...

Box or Battleship

In terms of structural integrity, you want your car to be somewhere between a cardboard box and a battleship.

It's no good to have a collision in a car that has the rigidity of a cardboard box. It will crumple very easily and absorb hardly any of the energy of the collision. Practically all the energy of the collision will be transferred to your frail, human body.

It's just as bad to have a collision in a car that has the strength and rigidity of a battleship. It will not crumple at all, and again, will absorb hardly any of the shock. Once again, practically all of the energy of the collision will be transferred to you.

What you want is a car that will crumple by absorbing practically all the energy. This will leave hardly any energy to damage you. The car will be wrecked, but you will live.

The difficulty for car engineers is that there are many different types of collisions — high speed or low speed, full head-on or just slight overlap head-on, single-vehicle rollovers, the front of one car running into the side of another car, etc. Even so, engineers have done a terrific job over the past 30 years and have saved many lives.

References

Allen, Shane, et al., 'A comprehensive statewide analysis of seatbelt non-use with injury and hospital admissions: new data, old problem', *Academic Emergency Medicine* (doi: 10.1197/j.aem.2005.11.003), 10 March 2006.

Ameratunga, Shanthi, et al., 'Road-traffic injuries: confronting disparities to address a global-health problem', *The Lancet*, 6 May 2006, pp 1533-1540.

http://www.snopes.com/autos/techno/seatbelt.asp

TOXIC CHOCOLATE

Theobroma cacao — the scientific name for the cacao tree from which we get chocolate — means 'food of the gods'. And chocolate lovers agree. People also love their pets. So should you share the love around and feed your pet some chocolate? Definitely not, because chocolate can kill pets, especially dogs.

Chocolate Tree

The cacao tree grows 6–12 m tall. After about four years, the mature tree begins to bear fruit, producing up to 70 pods, each about 35 cm long and ranging in colour from deep purple to bright yellow. Inside each pod are 20–60 beans, each 2–3 cm long — the Mother Lode of Chocolate.

Farmers remove the beans and bury them in large piles, so that they can ferment. During this process the flavour begins to develop. After a few days the beans are roasted. This reduces the water content, develops the flavour further and reduces the acidity and the bitterness. The beans are then popped out of their shells and ground into a paste called a 'cocoa mass' or 'chocolate liquor'. In the world of chocolate there is a special word called 'conching', which means to 'stir and aerate'. Depending on the quality, the gluggy, brown chocolaty mess is conched for between four and 72 hours, while being heated to between 55°C and 88°C. Sugar and milk may also be added at this stage. Once it has cooled down into 5 kg blocks, you have basic chocolate.

Baking chocolate and dark chocolate have the highest amounts (up to 85%) of pure cocoa paste and are usually bitter, because they have very little sugar added. Sweet chocolate has (of course) sugar added, as well as a whole bunch of other flavourings such as vanilla. To get milk chocolate, you add milk.

History

Chocolate was drunk by the Preclassic Maya some 2500 years ago. They frothed it into foam by repeatedly pouring the liquid from one vessel to another. By the time of the Spanish Conquest of Central and South America, chocolate was drunk with most meals, usually mixed with another ingredient such as water, honey or chilli. In 1502, Columbus brought back some cocoa beans to Spain. In 1528, Cortes brought three chests of cocoa beans to the Spanish King, Charles V. The Spaniards managed to keep the secret of their bitter but delicious drink for a century. Chocolate remained a drink until 1847, when J.S. Fry & Sons came up with a solid form of chocolate.

Pharmacology

Cacao contains over 500 different chemicals, including a few in the methylxanthine class. It has some caffeine (1,3,7-trimethylxanthine) and about seven times as much theobromine (3,7-dimethylxanthine). The theobromine seems to be the dog killer. (I don't know why human beings like methylxanthines, but we do.)

The toxic dose of theobromine is 100–150 mg/kg of body weight for dogs. Different types of chocolate have different amounts of theobromine. Milk chocolate has 154 mg of theobromine for each 100 g of chocolate. But for the same weight of chocolate, semi-sweet chocolate has 528 mg of theobromine (3–4 times more), while bitter baking chocolate carries a massive 1 365 mg of theobromine (nine times more).

One man's 'food of the gods' ... is another dog's death!

Feeding your pets chocolate (especially dogs) can kill them.

For your average 20–25 kg family dog, a fatal dose of chocolate is about 1.5 kg of milk chocolate, about 400 g of semi-sweet chocolate or just 140 g of cooking chocolate.

For humans, a fatal dose of milk chocolate is about 10 kg.

The average 'brown dog'

Bad Chocolate

For the average 20–25 kg family dog, the fatal dose of chocolate is about 1.5 kg of milk chocolate, about 400 g of semi-sweet chocolate — but only 140 g of dark chocolate. This is not a lot of chocolate for a dog to wolf down, especially when you consider that dogs don't really mind bitter tastes. (Cats are fussier about bitter tastes and so are much less likely to eat chocolate.)

For human beings, the fatal dose is about 10 kg of milk chocolate. We are saved from death by chocolate by three factors — our greater body weight, our lower metabolic rate and our slightly different biochemistry.

There is no antidote for theobromine, so a veterinarian can only treat the symptoms, which usually appear eight hours after eating the chocolate. The dog may suffer vomiting, diarrhoea, urinary incontinence, seizures or convulsions, blueish skin and a fast, irregular heart rate as it heads down theobromine's lethal pathway. The dog should be taken immediately to a veterinarian, who will try to stabilise the dog's breathing, heart rate, electrolytes, acid-base balance and overexcited nervous system. The veterinarian may also try to remove any remaining chocolate from the dog's gut with gastric lavage, activated charcoal or by inducing vomiting etc.

So yes, chocolate can indeed kill your dog.

Good Chocolate

However, in small doses, chocolate definitely appears to be good for human beings — especially dark chocolate. Cacao contains chemicals that fight the bacteria that cause tooth decay. On the other hand, milk chocolate is rich in sugar, which feeds these same bacteria. One of the good chemicals in chocolate goes under the obscure name of (—)-epicatechin.

One study looked at people with a long history of hypertension. Eating 100 g of dark chocolate every day for 15 days reduced their blood pressure by 11.9/8.5 mm Hg. It also improved their insulin resistance, reduced their LDLs (the bad cholesterol) and improved the dilation of their blood vessels.

Caffeine and Chocolate

The caffeine in coffee is a remarkably safe drug, but it can close down the blood vessels. Dark chocolate has the opposite effect.

So nowadays I try to have a small cube of dark chocolate about an hour before drinking coffee, to keep my blood vessels healthy. In the studies that looked specifically at chocolate opening up the blood vessels, the healthy volunteers took 100 g of dark chocolate, which also gave them one-quarter of their daily kilojoules. However, other studies show that eating only 10 g of dark chocolate has a good effect. In my case, even if the chocolate doesn't have the desired effect, it still tastes good. If you add chocolate to your diet to have with your coffee, make sure that you drop something else out of your daily diet, so that you don't overload on kilojoules.

Overall, the good chemicals in dark chocolate keep the blood vessels both elastic (which is good) and wide open (which is also good). Of course, this matches children's mouths, which are also elastic and wide open for chocolate …

Half-life

'Half-life' is the time taken to 'lose' half of what you started with. This term is commonly used with radioactive elements and drugs,

In a dog, theobromine has a half-life of about 17.5 hours. This means that if your dog eats too much chocolate, and loads up its blood with 100 mg of theobromine, after 17.5 hours the kidney and liver will have reduced it to 50 mg. After another 17.5 hours there'll be only 25 mg of theobromine remaining, and so on.

Chocolate, the Downer

Chocolate can have different effects on different types of people.

Some people are 'cravers', who love chocolate for the pure pleasure of it. They enjoy chocolate as they would a glass of fine wine. They feel good while eating the chocolate — and afterwards as well.

Then there are the 'emotional eaters', who eat chocolate to relieve boredom, stress or depression. They are looking for something (e.g. chocolate, food or alcohol) to make them feel better. If they have chosen chocolate, all they get is minor temporary relief (probably due to a full belly) — and after eating the chocolate, they feel worse. What a shame.

Onions Kill Dogs — and Cats

Chocolate is not the only human food that can kill our pets. The onions from your weekend barbecue are also potentially fatal.

It seems that the S-methylcysteine sulphoxide changes the red blood cells of our pets (but not their owners) to develop a structural defect known as a Heinz Body. The red blood cells are then taken out of circulation, making the pet anaemic. There has even been a case of a cat dying after eating onion soup.

References

Blumberg, Jeffrey, et al., 'Cocoa reduces blood pressure and insulin resistance and improves endothelium-dependent vasodilation in hypertensives', *Hypertension*, August 2005, pp 398-405.

Buijsse, Brian, et al, 'Cocoa intake, blood pressure and cardiovascular mortality: the Zutphen Elderly Study', *Archives of Internal Medicine*, 27 February 2006, pp 411-417.

Encyclopaedia Britannica, Ultimate Reference Suite DVD, 2006 - 'cocoa' and 'chocolate'.

Hurst, W. Jeffrey, et al., 'Cacao usage by the earliest Maya Civilization', *Nature*, 18 July 2002, pp 289, 290.

Parker, Gordon, Parker, Isabella and Brotchie, Heather, 'Mood state effects of chocolate', *Journal of Affective Disorders*, on line, accessed 20 March 2006.

Schroeter, Hagen, et al., '(—)-Epicatechin Mediates Beneficial Effects of Flavanol-Rich Cocoa on Vascular Function in Humans', Proceedings of the National Academy of Sciences, 24 January 2006, pp 1024-1029.

Vlachopoulos, C., et al., 'Dark Chocolate Improves Endothelial Function in Healthy Individuals', Abstract P638, 2004 European Society of Cardiology Congress, Munich.

'Why is chocolate bad for cats and dogs, but not for humans?' *Focus*, March 2004, p 48.

DREAM ON

My daughter Alice, who loves remembering and retelling her dreams, announced recently that she had a 'mythconception' for me. She had noticed that some of her schoolmates claimed that they did not dream.

There are many beliefs about dreams, that odd state of consciousness where we watch images flash by on the inside of our eyelids. We all have dreams, but remembering them depends on how soon we wake up after a dream. Statistics show that, on average, while we have several dreams every night, we remember a dream only about once a week.

Our fascination with dreams has led us to try to understand the nature of them.

Nature of Dreams 1: Reflect Reality

One interpretation of dreams suggests that they are not fantasy but an accurate reflection of reality. For example, in some tribes of Borneo, if a man dreams that his wife is unfaithful, then she is automatically assumed to be unfaithful in real life and has to leave his family. The *Encyclopaedia Britannica* reports that, 'a Macusi Indian of Guyana is reported to have become enraged at the European leader of an expedition, when he dreamed that the leader had made him haul a canoe up dangerous cataracts. He awoke exhausted, and could not be persuaded that the dream was not real.'

A Jesuit priest tells the story of an Iroquois Native American, who dreamed that 10 of his friends cut holes in the ice covering a lake and dived into one hole and up out of the other. The next day, the man told his friends of the dream. In accordance with their tribal law, they carried out the dream — but one of them drowned under the ice.

Nature of Dreams 2: Source of Divination

Others interpret dreams as a source of divination or knowledge about the future. The dreams of Joseph in the Bible have this kind of significance. This belief was common in ancient Greece, India and Babylon — and even today, if the dream books in the bookshops are any guide.

In fact, the oldest 'dream guide' dates back to 668–627 BC. This Babylonian tome was found in the ruins of the city of Nineveh, in the library of the Emperor Ashurbanipal.

The Roman emperor Nero dreamed that he was being covered in ants. He interpreted this to mean that his subjects would turn against him.

Dreams have also helped religious leaders. Mohammad (570–632 AD) gained insights into his path to Islam via dreams. But his subjects then practised this habit of dream divination so frequently that he forbade it.

Joseph Smith (1805–1844) founded Mormonism on the basis of his dreams. He said that an angel had told him how to find the buried golden tablets that showed that native Americans were actually direct descendants of the tribes of Israel.

Nature of Dreams 3: Curative

The ancient cultures of Greece, Babylon and Egypt used dreams as part of a mental or physical cure. It was called 'dream incubation'. Sick people would come to the temple to dream. The priests and priestesses would then interpret these dreams and give advice based on these interpretations.

More recently, Freud, who helped found psychiatry, said that 'dreams are the royal road to the unconscious'.

Nature of Dreams 4: Creative

Many individuals have used dreams for creative or intellectual inspiration, or for solving problems.

Samuel Taylor Coleridge wrote *Kubla Khan* as a result of a dream, while under the influence of opium. The German chemist F.A. Kekulé von Stradonitz is said to have worked out the chemical structure of benzene one night in 1865, while dreaming of snakes swallowing their own tails. He realised that the carbon atoms in benzene were not arranged in a straight line, but in a circle. And Robert Louis Stevenson said that the 'little people' who came to him in his dreams gave him inspiration.

Dreams Are Ordinary

Most dreams involve surprisingly boring topics (such as shopping or going for a swim). However, they seem weird because of the sequencing (e.g. you walk out of a shop and straight into the surf).

Dreams rarely happen in strange and exotic locations. One American midwestern university student had a very long (and boring) dream in which he mowed the lawn. The only objects visible in the dream were the grass, the lawnmower and his forearms controlling the lawnmower.

In most cases, dreams are fairly egocentric. The dreamer is part of the dream and, in most cases, knows the other people in the dream — family members about 20% of the time. Men dream of other men about 60–70% of the time.

REM Breakthrough

A major breakthrough into understanding dreams came with Drs Eugene Aserinsky and Nathaniel Kleitman in 1953. They first described REM sleep, which is when most dreams happen.

REM stands for Rapid Eye Movement. If you quietly sneak up on somebody during their REM sleep, you can detect their eyes flicking rapidly left and right under their closed eyelids, as though they are watching an invisible high-speed tennis match.

Today, sleep scientists tell us that we have three states of consciousness — awakefulness, REM sleep and non-REM sleep. (People who practise Transcendental Meditation claim that there is a fourth state — transcendental consciousness, or restful alertness.)

It is important to realise that both the REM and non-REM stages of sleep are active stages in their own right. The brain is not just having a rest and doing nothing. Sleep does not mean that the brain is doing nothing. Sleep is important reorganisation of neuronal activity.

In a typical night, we have four or five sleep cycles, each one about 90 minutes long. In the first cycle, we drift from a light sleep (Stage I) into a deeper sleep (Stage II) and then into Deep Sleep Stages III and IV. Stages I–IV are periods of non-REM sleep. We then go into reverse and come practically all the way back to awakefulness. But instead of waking, we enter REM sleep.

REM sleep is a strange form of consciousness. The brain is metabolically and electrically active, but in a way that is very different from its normal activity. It makes sense that the body is almost fully paralysed (apart from the flicking eyeballs), because this will stop you from acting out your dreams and hurting yourself or others. Your breathing, pulse and blood pressure are irregular, and lots of blood gets shunted to your sex organs. Men will usually have an erection during REM sleep, while women have clitoral engorgement.

During REM sleep, you will usually dream and, indeed, most dreaming happens during this time.

Then you dive down into the deeper stages of sleep again and, after another 90 minutes or so, emerge into your second REM

I dream, therefore I sleep ... or vice versa

Dreams have long fascinated us ... why do we remember some and not others, what do they mean, and why are they so spiggin weird?

Your sleep goes a little something like this

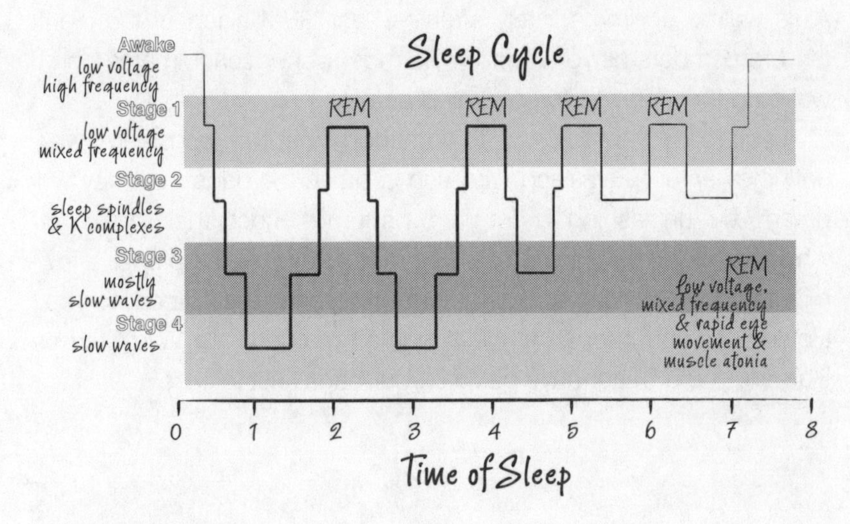

In a typical night, we have four or five sleep cycles, each about 90 minutes long.

Stages drift from light sleep (stage 1), to deeper sleep (stage 2) and then into deep sleep stages 3 & 4.

Stages 1 to 4 are non-REM sleep. We then reverse and come practically all the way back to awakefullness – but instead of waking, we enter REM sleep.

sleep of the night — and you'll dream again. You'll have four or five REM sleeps (and dreams) during an average night, each one being longer than the one before it. Indeed, about 20% of your night's sleep is spent in REM sleep, mostly dreaming.

Remember Dreams?

So, what about remembering your dreams?

If you are woken *during* a REM sleep, you will usually be able to describe your dream. But there is a very narrow window of memory. So, if you are woken 15 minutes *after* your REM sleep you usually won't remember your dream. So I'm guessing that my daughter Alice wakes up immediately after her last REM sleep of the night (and remembers her dreams), while most of her schoolmates don't wake up then (and so forget their dreams).

If you really want to prove to somebody that they do dream, wait until they enter REM sleep. You can improve the odds that they will dream by unleashing a fragrant perfume under their nose, or whistling, or blowing air across their face. Wake them up after a few minutes of REM sleep and ask them to describe their dream to you. However, if you happen to be a woman married to a Wild Man of Borneo, you'd better have packed your bags, just in case …

Reality

Sometimes dreams are so real that when we wake up from them, we are truly confused about reality for a few moments.

The English scientist and philosopher Bertrand Russell wrote: 'It is obviously possible that what we call waking life may be only an unusual and persistent nightmare. I do not believe that I am now dreaming, but I cannot prove that I am not.'

Never Dream

Every animal studied by scientists dreams — except for the Echidna. (In other words, the Echidna doesn't have any REM sleep.)

We're not sure why, but it might be related to the fact that the Echidna has no reticular system. The reticular system is part of the brain that, among other functions, 'filters' information coming into the brain. For example, at one particular instant, your brain might be receiving information that it's daytime, that you're playing Scrabble with your friends and that you might be able to lay down all seven of your tiles on the next go, that the coffee percolator is about to pop, and that the phone has just begun to ring. Your reticular system will help you sort out which are the most important pieces of information to deal with.

On the other hand, the Echidna has the largest pre-frontal cortex (in relation to its size) of any known animal, including human beings.

Perhaps without a reticular system the Echidna could not process the day's information while sleeping. Perhaps it had to evolve a huge pre-frontal cortex, to process the incoming information in real time. This pre-frontal cortex was fine in the small Echidna, but it would have been too big for the skull in more sophisticated animals. And so (perhaps), the reticular system in human beings evolved to allow the pre-frontal cortex to do other stuff — like write poetry, do income tax returns and make weapons of mass destruction.

REM = Dreams

In one study, people who were woken during REM sleep remembered their dreams vividly 20 times out of 27.

On the other hand, if they were woken during non-REM sleep, they remembered their dreams only four times out of 23.

How Much Sleep?

In higher animals, sleep is related to bigger brains. An animal sleeps if it can afford to. A lion spends most of the day sleeping. A smaller animal, such as a rabbit, spends more time awake, being alert for predators.

Sleep time is also related to diet. Carnivores sleep the most, herbivores the least, with omnivores somewhere in between.

References

Bodanis, David, *The Body Book: A Fantastic Voyage to the World Within*, London: Little Brown & Co, 1984, pp 252, 274-286.

Encyclopaedia Britannica, Ultimate Reference Suite DVD, 2006 — 'dream'.

Hobson, J. Allan, 'Sleep is of the brain, by the brain, and for the brain', *Nature*, 27 October 2005, pp 1254-1256.

Schatzman, Morton, 'Solve your problems in your sleep', *New Scientist*, 9 June 1983, pp 692-693.

Siegal, Jerome M., 'Clues to the function of mammalian sleep', *Nature*, 27 October 2005, pp 1264-1271.

Winson, Jonathan, 'The meaning of dreams', *Scientific American*, November 1990, pp 42-48.

OSTRICH HEAD IN SAND

The English language is very rich and descriptive. For example, someone 'burying their head in the sand, like an Ostrich' is foolishly ignoring their problem while hoping that it will magically vanish. But this evocative and widely used expression is not based on fact. The Ostrich does many things, but burying its head in the sand is not one of them.

The Ostrich

The Ostrich, found only in parts of Africa, is the largest living bird. It can measure up to 2.4 m in height and weigh 155 kg. If scared, it can run at speeds of up to 65 kph. Its kick is powerful enough to bend 10 mm steel rods into right angles and can easily break a human leg.

The Ostrich uses its wings for balance (when running) and for courtship and display. (By the way, its eyes have thick black eyelashes.)

It is mentioned, none too fondly, in one of the oldest books of the Bible — *Job* 39: 13–19:

'For she leaveth her eggs on the earth,
And warmeth them in the dust,

And forgetteth that the foot may crush them, Or that the wild beast may trample them.
She dealeth hardly with her young ones, as if they were not hers: Though her labour be in vain, she is without fear:
Because God hath deprived her of wisdom, Neither hath he imparted to her understanding.
What time she lifteth up herself on high, She scorneth the horse and his rider.'

This bad biblical rap doesn't stop ostrich farmers being attracted to the Ostrich's durable leather, beautiful saleable feathers, lean meat and extremely high feed-to-weight-gain ratio (3.5:1, much better than cattle at 6:1).

Ostriches have three main strategies when attacked. They can run away, kick or try to hide (e.g. when guarding their eggs). When hiding, they will sometimes lie flat on the ground, with their long neck and head also on the ground. In the wavy, hot air of their native Africa, they look just like a grassy mound.

Blame Pliny

The myth that an Ostrich will bury its head in the sand in an effort to hide may have begun with Pliny the Elder (23–79 AD), whose real name was Gaius Plinius Secundus. Pliny was a man of intense curiosity about the world around him. His nephew, Pliny the Younger, wrote about him, 'He began to work long before daybreak. He read nothing without making extracts; he used even to say that there was no book so bad as not to contain something of value. In the country it was only the time when he was actually in his bath that was exempted from study. When travelling, as though freed from every other care, he devoted himself to study alone. In short, he deemed all time wasted that was not employed in study.'

In 79 AD, Mt Vesuvius erupted, covering and thus preserving the city of Pompeii. While most people ran away from the Vesuvius eruption, Pliny went straight into the danger zone to look, learn and rescue survivors — and died in the attempt. In his honour, the most violent volcanic eruptions (such as Krakatoa) are called 'ultra-plinian'.

To bury one's head in the sand ...

'Hiding your head in the sand, like an Ostrich' is a saying that relates to foolishly ignoring a problem and hoping it will magically vanish.

The foolish thing is, Ostriches don't stick their heads in holes in the ground.

They are much more likely to run, kick or hide.

A dramatisation of the myth

Before his death, Pliny had almost completed one of the earliest comprehensive encyclopaediae. His *Natural History*, in 37 books, was a remarkable attempt to summarise all the knowledge known to the Romans. He claimed to have covered some 20 000 topics, partly using information from some 2000 books written by some 100 authors. In fact, he was one of the first writers to acknowledge the authors he quotes, and was also one of the first to provide a table of contents. His *Natural History* remained a fundamental source of knowledge to the West through the Dark Ages.

In Book 10, Chapter 1, Pliny writes of Ostriches: '... they imagine, when they have thrust their head and neck into a bush, that the whole of their body is concealed.'

Historians have assumed that this single sentence is the root of the myth about Ostriches burying their head in the sand.

Ostrich Tricks

Ostriches have one interesting piece of behaviour that comes close to burying their head in the sand. When Ostriches feed, they sometimes lay their head flat on the ground to swallow sand and pebbles. (The hard grit helps them to grind their food.) From a distance, the Ostrich looks like it's burying its head in the sand.

Will you ever see an Ostrich with its head in the sand? Not naturally. On the globalgourmet.com home page, Claire and Monty Montgomery describe how they visited the Brandywine Ostrich Ranch in Hemet, California, to see and eat Ostrich. The owner, Chip Polvoorde, told them how he helped get an Ostrich's head into a hole in the ground for a movie shoot. Chip's friend first dug the hole, laced it with yummy ostrich food and, once the unsuspecting bird had shoved its head into the hole, held it there with sheer brute force until the camera crew were happy. Some people will do anything for the camera ...

Pliny the Teacher

Pliny was a well-educated man from a wealthy family. In the army, he served in Germany, rising to Cavalry Commander. He then studied law in Rome and rose to Procurator of Spain. Back in Rome again, he had various official positions, partly because his friend Vespasian, who served with him in Germany, became Emperor.

(Useful advice: If you have a choice of an enemy in high places, or a friend in high places, go for the friend. Things magically become so much simpler.)

When he died, his position was Commander of the Fleet in Naples.

But Pliny's true love was learning and passing on what he had learned.

In his encyclopaedia of 37 books, Book I summarised the other 36 books. Book II covered astronomy and cosmology. Books III–VI discussed the geography of his world. Books VII–XI discussed zoology, from human beings to insects. Unfortunately, Pliny did not actually gather all this data. He simply recycled much of what Aristotle had written, and then added a few legendary animals and fantasy stories.

The next series, Books XII–XIX, dealt with plants. Here he combined stories gleaned from the past together with his own current observations. In Book XVII, he describes a grain harvester driven by oxen. This was long thought to be fiction, until an illustration of one was found in Belgium in 1958.

The major part of his work, Books XX–XXXII, deals with drugs and medicine. Finally, Books XXXIII–XXXVII cover precious stones, minerals, and how to turn them into metals.

Pliny's Exact Words

In Book X, *The Natural History of Birds*, Pliny wrote about the Ostrich. I was embraced by a sense of awe as I read these 2000-year-old words.

'The history of the birds follows next, the very largest of which, and indeed almost approaching to the nature of quadrupeds, is the Ostrich of Africa or Ethiopia. This bird exceeds in height a man sitting on horseback, and can surpass him in swiftness, as wings have been given to aid it in running; in other respects Ostriches cannot be considered as birds, and do not raise themselves from the earth. They have cloven talons, very similar to the hoof of the stag; with these they fight, and they also employ them in seizing stones for the purpose of throwing at those who pursue them. They have the marvellous property of being able to digest every substance without distinction, but their stupidity is no less remarkable; for although the rest of their body is so large, they imagine, when they have thrust their head and neck into a bush, that the whole of the body is concealed. Their eggs are prized on account of their large size, and are employed as vessels for certain purposes, while the feathers of the wing and tail are used as ornaments for the crest and helmet of the warrior.'

References

Barham, Andrea, *The Pedant's Revolt: Why Most Things You Think Are Right Are Wrong*, London: Michael O'Mara Books, 2005, p 31.

Collins, Barry, 'The Belfast Agreement: the purloined letter and the "Politics of the Ostrich"', *International Journal for the Semiotics of Law*, 2002, Vol 15, pp 273-292.

Exploring the Secrets of Nature, London: Reader's Digest Association Limited, 1994, pp 20, 31, 52, 149, 192, 222.

Facts and Fallacies: Stories of the Strange and Unusual, Sydney: Reader's Digest Association, Inc., 1988, p 52.

http://www.globalgourmet.com/food/egg/egg0197/ostrich.html

Nash, David. S., 'The failed and postponed millennium: secular millennialism since the Enlightenment', *The Journal of Religious Study*, February 2000, pp 70-86.

Parejko, Ken, 'Pliny the Elder: rampant credulist, rational skeptic, or both?', *Skeptical Enquirer*, 1 January 2003.

WATER RECYCLING, WATER RECYCLING, WATER RECYCLING

Australia is the driest inhabited continent on Earth. About 80% of its surface doesn't have enough rain to grow crops — and half of this is classified as desert (i.e. less than 200 mm of rain per year). For this reason, it's common for parts of Australia to have the occasional water crisis. Sydney's main dam, Warragamba, has dipped to less than 40% full. Sydney also has the lowest recycling figure (around 3%) of any Australian city. Perth and Brisbane recycle about 4%, while Adelaide recycles 11%. Canberra, Melbourne and Perth have set targets to recycle about 20% of their water by the year 2012. These figures are embarrassingly far below what many European cities routinely achieve.

However, some Australians are opposed to any recycling whatsoever of their drinking water. They are deeply disturbed by the thought of drinking recycled sewage. But I really don't understand their thinking. The truth is that virtually all of the water we have ever drunk has been recycled many many times.

'Pure' drinking water really doesn't fall magically from Heaven to be used only once and then sent to some unknown Hell, from which it never returns. Over millions of years, the water we drink has been through the digestive system of many animals, from sharks to dinosaurs to fish.

Water — the Molecule

Water is probably the only chemical for which most people know the formula — H_2O. This formula tells us that two atoms of hydrogen are married to one atom of oxygen. The molecule looks like a fat 'V' or a tiny boomerang, with the oxygen atom in the middle, the hydrogen atoms at the ends, and the angle of the boomerang set at 104.5° (just a little bigger than a right angle, which is 90°).

Make Your Own Water

If you are really keen, you can make your own drinking water. One way is to mix hydrogen gas and oxygen gas, and set them alight with a match. You get a bang, lots of energy and the waste product called 'water'. (Don't do this at home, folks — it can be dangerous!) The chemical reaction releases a huge amount of energy.

In fact, this same chemical reaction helps push the space shuttle into orbit. At liftoff, the shuttle's external tank holds over 100 tonnes of liquid hydrogen and 600 tonnes of liquid oxygen. (Hydrogen has the lowest density of any element. That's why 100 tonnes of liquid hydrogen has twice as many atoms as 600 tonnes of liquid oxygen, even though it has one-sixth of the mass.)

By the way, Fuel Cell cars, the next generation of cars, also use this chemical reaction. They burn hydrogen with oxygen to make energy and water. The big difference is that the space shuttle burns the hydrogen with oxygen as quickly as possible to get the maximum amount of energy, while the Fuel Cell car burns it as slowly as possible to get maximum fuel economy.

Another way to make water is as a by-product of various chemical reactions. Combining chemical A with chemical B produces chemical C plus water.

But these are really uncommon ways to get your water.

Most of our water has a very ancient history.

Water From Space

Yes, it's true. More than 99.99999+% of the water we drink today was made a very long time ago and very far away as well. We have to go back to the beginning of the Universe.

The Big Bang happened about 13.7 billion years ago. After about 400 000 years, the early Universe had cooled down enough to make the first hydrogen atoms. After about 400 million years, the first generation of stars sparked into existence and as part of their nuclear burning made oxygen. At a later stage in their life cycle, these stars hurled some of their substance into the space around them. Some of this star stuff was oxygen atoms. When one of these oxygen atoms combined with two hydrogen atoms, the result was a single molecule of water. So water molecules were probably first created 10–12 billion years ago. Soon there were huge numbers of water molecules very loosely held together by gravity. These clouds of molecules of water were gravitationally attracted to clouds of other molecules (e.g. hydrogen, alcohol, water, etc.). The clouds eventually coalesced to form a new generation of stars, some of which had planets around them.

As an aside, stars are continually forming and exploding, so new water molecules are being made in space all the time. But very few of them reach our planet.

Water From Earth

Our planet, Earth, is much younger than its water.

Earth formed about 4.6 billion years ago. In its early days, it was a ball of red-hot molten lava, thanks to the heat generated from the impacts of in-falling comets, asteroids, planets the size of Mars, etc. After the bombardment, the planet began to cool. Over 50 million years or so, the heavier elements (such as iron) sank to the centre, while the lighter elements floated on the surface. The volatile chemicals (such as water, carbon dioxide and nitrogen) dissolved in the molten lava on the surface.

As the lava cooled, it released the water and other lighter chemicals into the atmosphere. But the temperature was still

It leaks water everywhere

The water on Earth goes around and around, being continually recycled. This is called the 'Hydrological Cycle'. This recycling is powered by the sun.

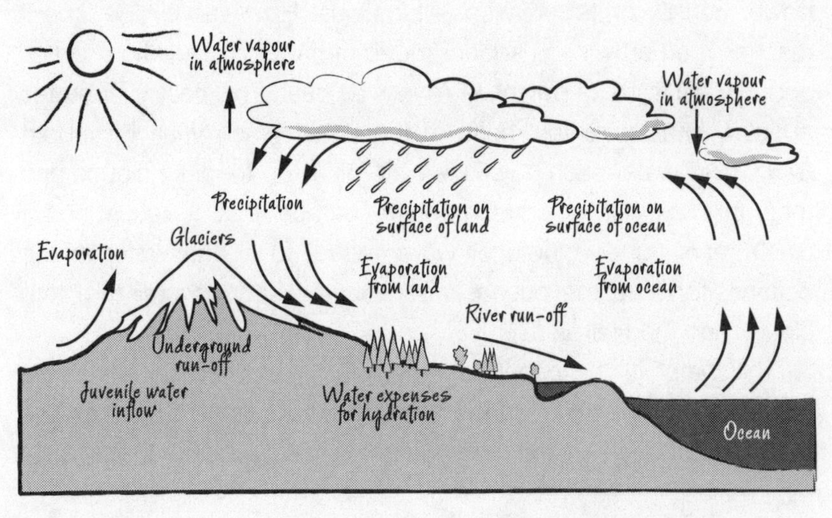

The much talked about 'Hydrological Cycle'

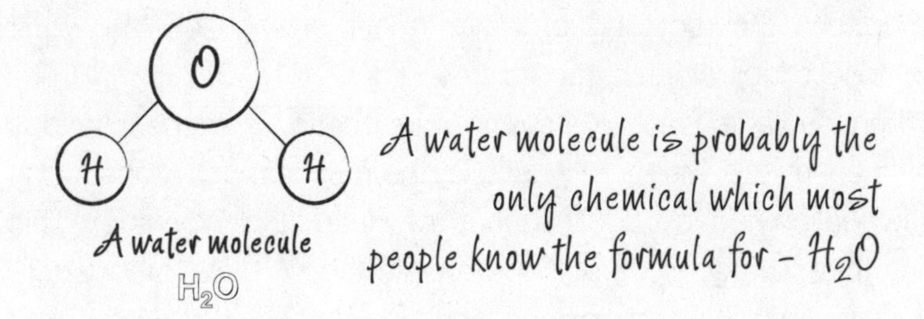

A water molecule is probably the only chemical which most people know the formula for – H_2O

fiendishly hot. Somewhere between 4 and 4.4 billion years ago, the temperature cooled enough for the steam to condense into liquid water and soon clouds appeared, followed by the first rains. It was about four billion years ago that molecules of liquid water appeared on the Earth's surface.

Hydrological Cycle

The water on Earth is being continually recycled. Geographers call it the Hydrological Cycle.

This recycling is powered by the Sun. Close to the Sun, the power density of its radiation is ferocious. However, by the time it reaches the Earth, each square metre of the Sun's radiation carries about 1000 watts of power — mostly as heat. This heat evaporates about a one-metre thickness of the ocean each year, i.e. about 875 km^3 of water each day. This water rises up to make clouds and then falls as rain. Oceanographers tell us that it takes about 3100 years for a volume of water, equal to all the water in the oceans, to leave the oceans, rise into the atmosphere, turn into clouds and then fall as rain.

The water in the oceans, rivers and lakes is continually being recycled. Some newly made molecules of water do enter the Hydrological Cycle — but their number is microscopic.

(Practically) All Water is Recycled

It's fairly easy to recycle bath and kitchen water. However, people who oppose the recycling of water quote sewage as the worst case, so let's look at sewage.

When sewage water runs into the ocean, it carries various organic chemicals with it. In Australia, the ratio is roughly one tonne of sewage carried by 2000 tonnes of clean drinking water. This means that sewage is 99.95% pure water and 0.05% 'other stuff'. The 'other stuff' is a mix of chemicals that dissolve in water, and solids that don't dissolve in water.

The heat of the Sun heats up the water, which makes the tiny boomerangs of individual water molecules move faster. Some of them move fast enough to break loose from the other water molecules and evaporate into the atmosphere. The water molecules rise from the ocean to form clouds and then rain.

When they rise into the air, they leave behind all the organic molecules of the 'other stuff'. So the heat of the Sun separates the water from the impurities. Once separated, each water molecule has no memory that it was once in contact with 'other stuff'. Each water molecule is, once again, as 'pure' as it was when it was first assembled from individual atoms of hydrogen and oxygen.

It doesn't matter whether the water molecules are separated from their organic friends by electricity in a recycling plant or by the heat of the Sun. The point is that the water molecules are being recycled again, as they have been many times before over many billions of years, and will be recycled far into the future. Many creatures have drunk your drinking water, long before you got to it.

Leonardo da Vinci, the great all-rounder from the Renaissance period, recognised this when he wrote, '... the heat of the Sun calls up ... moisture from the expanses of the sea ... the waters pass from the rivers to the sea, and from the sea to the rivers, ever making the self-same round, and ... all the seas and the rivers have passed through the mouth of the Nile an infinite number of times ...'

One Kilowatt Per Square Metre?

If you take a cross-section of the Sun's radiation, you will find that it carries about one kilowatt of heat. And yes, it does indeed shower about one kilowatt of heat onto a square metre of the Earth's surface — but only if the Sun is directly above this part of the Earth's surface, say, at the equator. If this square metre of the Earth's surface is far from the equator, say inside the Arctic Circle, then this one square metre of the Sun's radiation gets spread over (say) 10 m^2 of the Earth's surface. This is part of the reason why it's so cold at the Poles.

A Few Water Numbers

Australia has an average rainfall of 469 mm per year compared to the world average of 746 mm. Yet Australia has the third highest use of water — 300 000 litres per person per year.

In Sydney, the average household produces 586 litres of waste water each day. Most of this is grey water (non-sewage, e.g. from the washing machine, shower and sinks) that could easily be recycled. On average, only 3–4% of Sydney water is recycled.

London recycles 80% of its water.

Right and Wrong Names

The gas hydrogen can be burnt with oxygen to make water. The name 'hydrogen' literally means 'water former'.

About 70% of the surface of our planet is water, so our home should really be called Planet Water, not Earth. In fact, the Pacific Ocean by itself is bigger than all the continents put together.

I attended a lecture by John Young, who piloted the space shuttle on its very first flight. (He was also one of the 12 men who walked on the Moon.) He made a joke about landing back on Earth after the Apollo 16 mission: 'We aimed to land in the Pacific Ocean, which is bigger than all the land masses on Earth, and we figured that was close enough for government work.'

References

Ball, Philip, *H_2O: A Biography of Water*, London: Weidenfeld & Nicolson, 1999, pp 4-27.

IDENTICAL FINGERPRINTS?

Anyone who has watched any of the current genre of forensic science TV dramas will know of the latest electronic ways to track down baddies. Even those of us who haven't kept up with the latest trends would have heard of good old-fashioned, reliable fingerprinting. However, it is a myth that fingerprint identification is infallible.

Anatomy of Fingerprints

Fingerprints begin to develop in the uterus when the foetus is about ten weeks of age. The deeper basal layer in the skin of the fingertips begins to grow faster than the upper layers, causing stress in the basal layer so that it buckles. This generates the well-known fingerprint ridges on the surface.

Your skin consists of a number of different layers. But in the hard-working areas which are used for gripping, some of the deeper layers have 'corrugations' (hills and valleys), so that they won't slip over each other. These corrugations that 'zipper' into each other are just below the surface. They show their existence on the surface as fingerprints (and toe prints and palm prints). These prints increase friction, giving you a better grip.

History of Fingerprints

Fingerprints have been used for identification for thousands of years. The ancient Chinese and Assyrians used fingerprints on legal documents. The ancient kings of Babylon would press their entire right hand into a slab of wet clay carrying a decree. Then they would fire the clay, thus preserving the decree for thousands of years. Even then, it was a common belief that no two people shared the same fingerprints.

Thomas Bewick (1753–1828), a master engraver of wood blocks, used his thumbprint to identify his work. He obviously believed that fingerprints were unique.

However, Henry Faulds, a Scottish doctor working in Japan, became the first person to solve a crime using fingerprints. He matched the fingerprints found on a cup at a robbery in Tokyo with those of a servant and described his feat in a letter to the prestigious science journal *Nature* in 1880.

In 1888, Sir Francis Galton, who had read Faulds' letter in *Nature*, realised that fingerprints were a unique way of identifying criminals. Galton began to study fingerprints. He noticed that fingerprint ridges would sometimes suddenly split or end, and that there were about 35–50 such features on an average finger. He noticed that these features, which he called 'minutiae', were different in all the fingerprints that he examined. He therefore concluded that the fingerprints of everyone in the world were different. He calculated that the chance of any two fingerprints being identical in all 35–50 minutiae to be about one in 64 billion (roughly the number of human fingers in the world today).

Matching Fingerprints

There are four basic patterns used to classify fingerprints — the arch, the loop, the whorl and the composite (which is a combination of the first three patterns). If you think of a military map or a bushwalker's map, an 'arch' is a gentle rise, a 'loop' is a ridge, while a 'whorl' is a solitary hill or peak.

Read these!

Fingerprints begin to develop when we are about 10 weeks old in the uterus. They increase our grip and, in close-up, look like a military map with hills, ridges and valleys.

The deeper basal layer in the skin of the fingertips grows faster than the upper layers. This causes stress in the basel layer so it buckles, which generates the well-known fingerprint ridges on the surface.

To match one fingerprint to another, the fingerprint officer has to find a certain number of recognisable features that are common to each fingerprint. These are called 'minutiae', 'ridge characteristics', 'Galton details' and 'points of similarity'. Police departments throughout the world use between eight and 16 matching points of similarity to judge whether fingerprints are 'identical'. For example, the Spanish police require 10 points of similarity to verify that two fingerprints are identical, while the British police require 16.

If you want to go deeper, you can examine the fingerprint lines themselves — their thickness, their separation, the depth of the valleys between the lines, and so on.

Fingerprints Are Not Identical?

So why isn't fingerprint identification foolproof?

First, no-one has ever proved that all fingerprints are different — it's just a claim that has been made over and over again. Similarly, it has always been stated that all snowflakes are different. But in 1988, the scientist Nancy Knight found two identical snowflakes in a Wisconsin snowstorm (see page 148).

Second, there's the human element. Human beings aren't perfect — we make mistakes (we're only human after all). The Amish people deliberately incorporate a mistake into every quilt they make — to show that only God is perfect. Every blood test done by a pathologist comes with an estimate of the error, as does every measurement made by a scientist. Pregnancy testing has an error rate, as does HIV testing. Why is it that, of all the measurements made by human beings, the fingerprint matching test is the only one that is claimed to be free of error? Even the US Department of Justice believes that fingerprint identification has a 'zero error rate'.

Third, Simon Cole, Assistant Professor of Criminology at the University of California at Irvine, has found 22 cases in which errors in fingerprint identification led to innocent people being convicted. Disturbingly, most of the mistakes were uncovered only as a result of exceptional circumstances, such as DNA testing of 10-year-old samples or unexpected confessions. Overall, he looked at a survey of proficiency of fingerprint laboratories, with regard to positive identification of fingerprints. Their average error rate was 0.8% — and one laboratory got 4.4% wrong. An error rate of 0.8% means that eight out of every 1000 fingerprint identifications are wrong. Such an error rate could set free eight wrongdoers or jail eight innocent people in every 1000. In the USA, this error rate equates to 1900 mistaken fingerprint matches in 2002 alone!

The Official Line

At the moment, the official pronouncement of the International Association for Identification (the world's largest and oldest forensic association) is that any fingerprint expert giving 'testimony of *possible* or *likely* (fingerprint) identification [as opposed to *definite* identification] shall be deemed to be engaged in conduct unbecoming'. In other words, the official line is that they never admit to any possibility of error.

This is a shame. Fingerprint identification is a powerful tool and should remain part of forensic science. But it should admit to having a small error rate — like every other measurement ever taken by human beings. Cole does admit that fingerprints may be unique when the entire fingerprints are examined, but not when only a small section of one finger is checked.

Otherwise, it may be discredited by judges and juries, who should be pointing the finger at the actual error rate of the fingerprint scientists.

A Typical Case of Fingerprint Mis-identification

On 11 March 2004, terrorists bombed the Madrid train network killing 191 people. After the bombing, Brandon Mayfield, a convert to Islam and an attorney in Oregon, was held for two weeks as a 'material witness'. However, he did not hold a passport and claimed that he had not left the USA in the previous 10 years.

He was implicated solely because of a fingerprint found on a bag in Madrid after the bombing. The bag was found to contain explosives and detonators. The Spanish National Police could not identify the fingerprint, and emailed it to other agencies around the world. In the USA, FBI Senior Fingerprint Examiner, Terry Green, identified it as belonging to Brandon Mayfield. The US Government's affidavit stated that Green 'considers the match to be a 100% identification' of Mayfield. This was supported by other fingerprint examiners, some with 30 years' experience.

Two weeks later, the Spanish National Police announced that the fingerprint belonged to Ouhnane Daoud, an Algerian living in Spain. The FBI retracted the identification and issued an apology to Mayfield.

The unusual feature in this case was that the fingerprint mis-identification was found *before* the accused was found guilty and sent to jail.

Another Fingerprint Mis-identification

Stephen Cowans was sent to jail for 30–45 years for shooting and wounding a police officer. He was convicted on fingerprint and eyewitness evidence.

In January 2004, after serving six-and-a-half years of his sentence, he was freed. Later, DNA testing showed that he was not the perpetrator. The Boston Police Department then admitted that the fingerprint evidence was wrong. (And often eyewitness evidence can also be wrong.)

As in the Brandon Mayfield case, there were many wise and experienced experts, who all testified at the time that the fingerprint evidence definitely pointed to just one man, the accused.

References

Coghlan, Andy, 'Fingerprint evidence stands accused', *New Scientist*, 31 January 2004.

Coghlan, Andy, 'How far should prints be trusted', *New Scientist*, 17 September 2005, pp 6, 7.

Cole, Simon A., 'More than zero: accounting for error in latent fingerprint identification', *The Journal of Criminal Law and Criminology*, 2005, Vol 95, No 3, pp 985-1078.

McKie, Iain, 'Prints are fallible', *New Scientist*, 8 October 2005, p 26.

'The Myth of Fingerprints: pretending evidence is infallible does no one any good', *New Scientist*, 17 September 2005, p 3.

FINGERPRINTS OF TWINS

A while ago, I test-drove an Audi A8 for a week (one of my other jobs). Many of its features impressed me, but the 'gee-whizz' feature was the Fingerprint Recognition Pad. I would rest the fingerprint of my little finger on a tiny glass window near the gearstick. Once the car recognised me, it would adjust the seat, mirrors, engine and gearbox, etc., to suit me. One of my passengers said, 'But it would be fooled if you had an identical twin, wouldn't it?'

He, like so many other people, wrongly believed, that identical twins have identical fingerprints.

Identical Twins

On average, about one in every 80 births results in twins. About one-third of these twins are identical, i.e. about one in every 250 births. Identical twins happen when a single fertilised egg splits into two embryos.

Each of the identical-twin embryos starts off with identical DNA. So shouldn't the newborn babies greet the world with identical fingerprints?

Maybe not so identical ...

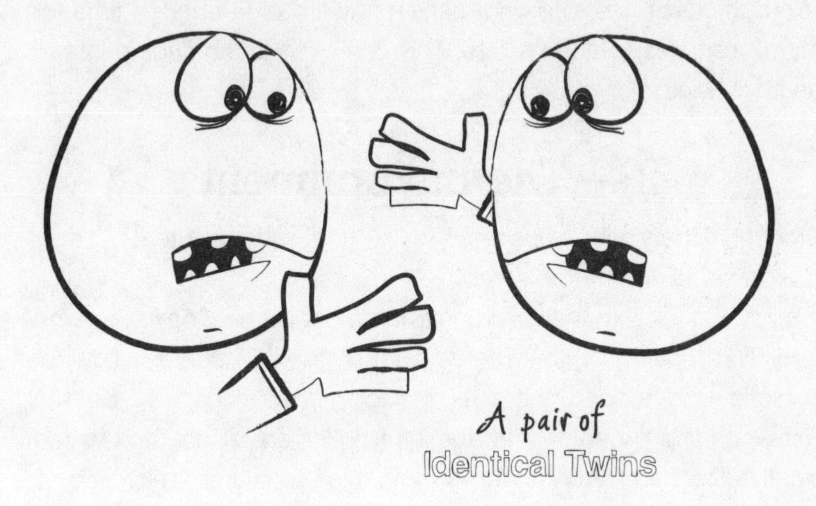

A pair of
Identical Twins

On average, around one in every 80 births results in twins. About one third of these are identical. Identical twins happen when a single fertilised egg splits into two embryos.

At least three factors influence the development of fingerprints, which are laid down between weeks 13 and 19 of gestation in the womb.

These three factors are the DNA itself, the environment and, strangely, how the environment acts on the DNA.

1 — The DNA

Very early in life, there is a single fertilised egg. This egg splits into two identical copies, which then grow into two identical babies.

However, the DNA doesn't make a perfect copy of itself every time it divides. Down on the micro scale, the DNA looks like a ladder a few metres long. To make it fit into the tiny cells of your body, this ladder is curved and twisted and bent. Your average household ladder might have a dozen rungs. The DNA ladder of life has about three billion rungs. Each time it divides or makes copies, mistakes can happen.

To get from one single cell to the 10–100 trillion cells of a human being, the DNA has to divide many many times — enough time for lots of mistakes to accumulate. This can change the fingerprints of identical twins.

2 — The Environment

Although identical twins grow in the same uterus, their environments are not the same.

First, in their external environment the babies are floating around in different parts of the uterus with a slightly different flow of amniotic fluid around them — including their fingertips. So the corresponding fingertips (say, the left little finger) will each grow in a slightly different environment. As the babies grow from microscopic dots to big babies, small differences will be preserved and amplified.

Second, in their internal environment the babies will almost always have different lengths and diameters of umbilical cord. This means that each twin will receive a slightly different amount of blood from the placenta. The twin who gets slightly less blood will make sure that its brain gets enough blood, even if this means stealing blood from the lower body. By an accident of anatomy, sending more blood to the brain automatically sends more blood to the arms. This means slightly bigger fingers, in the same way that if you blow more air into a balloon it gets bigger.

Mathematicians tell us that bigger fingertips mean more whorls.

If the foetus had flattened finger pads, the baby's fingerprints are more likely to have the simple arch or loop pattern. On the other hand, if the foetus had swollen finger pads, the fingerprints are more likely to have the more complex whorl pattern of ridges.

3 — Environment and DNA

Scientists have long wondered if the post-birth environment could act on the DNA. They have noticed that in a pair of identical twins, one might become a diabetic or suffer from schizophrenia or depression, while the other does not.

It was Dr Fraga (from the Spanish National Cancer Centre in Madrid) and his colleagues who first explained some of what was going on. They studied 40 pairs of twins, ranging in age from 3–74 years.

Twins start off with essentially the same DNA. But as they grow, chemicals act upon the DNA, changing how the DNA acts.

One chemical change is called 'methylation', i.e. methyl groups are added. Methylation acts like a brake, slowing down the activity of that section of DNA. So methylation could reduce the production of (say) growth hormone, making one twin shorter. Another chemical that can be added to the DNA is the acetyl group, leading to 'acetylation'. This change acts like an accelerator and could increase the production of growth hormone, making one twin taller.

These chemical changes can be brought on by the environment — diet, infectious diseases, cigarette smoke, exercise, etc.

Dr Fraga's team found that the younger the twins were, the more identical they were. On the other hand, the older they were, and the more time that they spent in different environments, the more different they were.

One well-known example is the agouti mouse. Here, the diet affects the methylation of a small part of the DNA (the 'inserted intracistenal A particle element', seeing as you asked). The result is that the mouse's coat changes colour.

The Experiment

Theory is always a good start. But you cannot fool nature — the experiment tells you the truth.

In 2002, Jain, Prabhakar and Pankanti did the experiment, publishing the results in their paper, 'On the similarity of identical twin fingerprints', in *Pattern Recognition* (The Journal of the Pattern Recognition Society).

The study found that the fingerprints of 94 identical twins were not identical, although they were more similar than those of non-related people.

The authors also looked at the example of an automatic fingerprint verification system, comparing the fingerprints of a million people. If the people were not related, the automatic system would wrongly accept 10 000 of them as being the same. But if they were twins, it would wrongly accept 48 000 of them as being the same. So again, twins' fingerprints were different from each other — but closer than those of complete strangers.

Various organisations collect fingerprints, but the FBI probably has the biggest collection. They have over 200 million sets of fingerprints — but then, so does every house with small children!

Biometric Verification

Various governments are trying to push automated biometric verification systems. 'Biometric' means 'measuring some physical or behavioural characteristic'. These systems need four properties.

First is 'universality'. Everybody must have it. Yes, we all have fingerprints.

Second is 'permanence'. Fingerprints barely change over your lifetime, except for injuries.

Third is 'collectibility'. Fingerprints are certainly easy to collect.

Fourth is 'distinctiveness'. Usually, everyone's fingerprints are different.

Whorls and High Blood Pressure

We're not exactly sure why, but if your fingerprints have lots of whorls, you are more likely to have high blood pressure. If the baby in the uterus has higher blood pressure in its arms, the fingerprints will grow more whorls.

Are these two facts related? We don't know.

Twin Fraud

Twins have been known to use their 'identicality' to trick teachers at school — and to even take exams for each other.

There have been adult criminal cases where twins have bought a single insurance policy to cover both of them, or where they have claimed unemployment benefits twice when only one of them was unemployed. There have even been cases where one twin was sent to jail for a crime committed by the other twin.

References

Fraga, Mario F., et al., 'Epigenetic differences arise during the lifetime of monozygotic twins', *Proceedings of the National Academy of Sciences*, 26 July 2005, Vol 102, No 30, 10604-10609.

Pankanti, Sharanth, et al., 'On the similarity of identical twin fingerprints', *Pattern Recognition*, Vol 35, November 2002, pp 2653-2663.

Wade, Nicholas, 'Explaining differences in twins', *The New York Times*, 5 July 2005.

Weiss, Rick, 'Twin data highlight genetic changes: minor differences increase with age', *The Washington Post*, 5 July 2005.

'Why do identical twins have different fingerprints?', *Focus*, June 1994, p 9.

LESS IS MOORE

DISCLAIMER
This story is partly about the mythconception that the rapid pace of change in computer technology has happened only very recently. This is not a widely held mythconception. However, I thought that it was such an important story that I had to crowbar it in somehow.

About 60 years ago, the electronic computer was a monster that filled three air-conditioned rooms, weighed 30 tonnes and chewed up enough electricity to run a village. It has now shrunk down to the modern battery-powered PDA (Personal Digital Assistant) that slips into your shirt pocket. Besides having a computer millions of times more powerful than the 60-year-old computers, the PDA also includes wireless and phone Internet access, a voice phone, a camera and a sound recorder. Most of us don't have a PDA but we probably use a laptop or a desktop computer. Computers have become very cheap, very powerful — and very small.

We have all become used to the relentless progress of computer technology. But most of us think that it has only been happening for a short while — five, perhaps 10 or 20 years. Actually, it's been happening for the past 60 years, and at pretty well the same breakneck speed each year.

History of Computers

In 1947, the ENIAC was probably the first digital stored-program electronic computer. For nine years, from 1947–1956, it was the most powerful electronic computer on Earth. If big is beautiful, ENIAC was drop dead gorgeous. It had 19 000 vacuum tubes and 1500 relays. It weighed 27 tonnes, filled about 460 m^3, and could add 5000 numbers every second. To do all of this, it chewed up 170 kW of power — enough to run a small village.

Two years later, in 1949, the March edition of *Popular Mechanics* reported that in the distant future computers might have 'only 1500 vacuum tubes and weigh only 1.5 tonnes', and still be as powerful as the ENIAC computer.

Gordon Moore

In 1965, Gordon Moore proclaimed what is now called Moore's Law, which basically states that every 18 months or so we shrink electronic components to half their previous size. In other words, we double the packing density of information. In 1969, Gordon Moore and a few friends set up the Intel Company.

Moore's Law applied from 1947–1965, and from 1965 to the present. We have moved from large vacuum tubes to little vacuum tubes to transistors to integrated circuits. Over the past 60 years, the shrinking time has varied between 12 and 24 months — but on average, Gordon Moore was remarkably correct.

Electronic components give you very good value for money. Over the past 60 years just about every known commodity — from the car you drive and the groceries you eat to the house you live in — has had a massive increase in price. Over the same time, computers have dropped in price — thanks to Moore's Law.

The Big Shrink

Various electronic components benefit from this shrinking.

First is the volatile working memory that stores information (RAM or Random Access Memory). My very first computer in 1987 had

1 MB (megabyte) of RAM. My current laptop has 2000 MB of RAM. My desktop computer has 8000 MB of RAM.

Second is the microprocessor that does the logic or thinking. My microprocessor used to amble on at a few MHz, but today it flashes by at 1670 MHz.

Third is the hard drive inside my laptop that stores my files and applications. It has expanded from 20 MB to 160 000 MB (or 160 GB, or gigabytes). (My stand-alone external hard drive, which I use to make an off-premises back-up of all our computers every month, has a capacity of 1 000 000 MB — 1 TB, or terabyte.)

And even though my current laptop computer is much more capable, it cost half the price of my desktop computer of 20 years ago.

Bad Software

If my computer is so much faster and cheaper, why does it still take a long time to start up?

One unexpected advantage of Moore's Law is that it helps really badly written software to keep on running. Software has been getting relentlessly bigger and 'buggier', with more and more bugs in it. Some software engineers call it 'bloatware', because it continually gets bigger.

Thanks to Moore's Law, the faster machines can deal with this crummy software, so that it still runs. Moore's Law has given the makers of badly designed software a free ride. Some computer software companies have the saying: 'Don't worry how big or bad it is, we'll just throw some faster hardware at it.'

Silicon Roadblock?

If we push Moore's Law to the year 2050, doubling the number of electronic components every 18 months, we would have a very powerful computer indeed. It would have the storage capacity of 200 000 human brains and a processing speed about 500 million times faster than the current computer chips.

However, Dr Muller and colleagues from the Bell Laboratories of

The bigger, the better

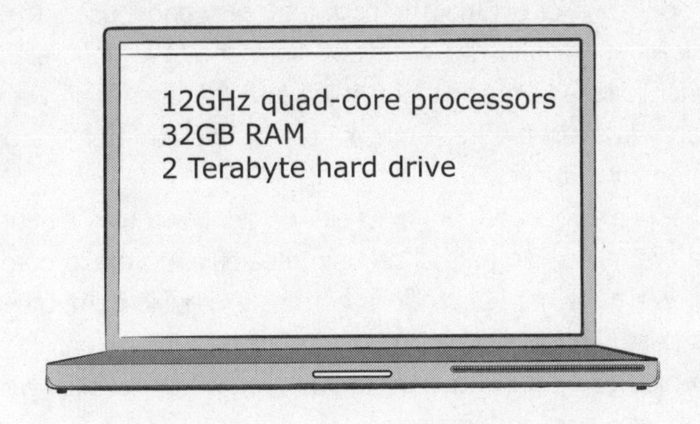

*Bigger, faster, badder, better looking, sexier, hotter, cooler and phat-er ...
AND it now holds 4 months of continuous music!*

Lucent Technologies in New Jersey reckon that we'll hit a major technological roadblock in Moore's Law in around 2012.

The average transistor is just an electronic switch. Electrons come in, come up against a 'gate' and at the right time they are allowed through. The gate is not a mechanical swinging device — it's made from silicon. It controls whether the electrons get through.

The fundamental problem is that the gate will get too thin. In 1997, a silicon gate was about 25 atoms thick. Following Moore's Law, by 2012, the gate in your transistor will be just five silicon atoms thick. Silicon only five atoms thick is a lousy insulator. Electrons will be able to leak through this thin layer, making the transistor unable to work reliably.

Therefore, if we want to make ever-smaller computer chips, we can't use *silicon*. We'll have to abandon silicon and try something else.

Moletronics

Engineers have a saying, 'Don't reinvent the wheel'.

Up until now, we've been making very small electronic circuits by getting big circuits and shrinking them down photographically. But why don't we copy the method that our body uses to make chemicals such as insulin or thyroid hormone? We don't start with big chemicals and break them down into the chemicals we want. Instead we assemble these chemicals by sticking atoms together, one atom at a time.

This has produced a new field called Molecular Electronics, where we try to use natural organic molecules to make computer chips. We have not yet made a complete organic computer chip, much less an organic computer. But so far scientists have assembled tiny organic molecules that behave like wires, switches, memory elements and transistors.

Abandoning silicon for organic chemicals might keep us on a Moore's Law pathway to ever smaller electronic components.

Big Jumps Sideways

Let's go beyond just improving or tweaking our current technology. Instead, let's make a complete jump into really weird, and very small and fast computers — while still following the Laws of Physics.

We have had a few of these big jumps in other fields. We didn't improve slide rules, we replaced them with electronic calculators. We didn't soup-up vacuum tubes, we replaced them with transistors and then integrated circuits. And we didn't fine-tune carbon paper, we abandoned it and invented the photocopier.

Let's look into the crystal ball to see what life could be like in 50 years' time.

There are three main possible future computers that have completely different technology from anything we have today. They are optical computers, DNA computers and quantum computers.

Optical Computers

We have been researching optical computing for about 25 years. We already use optic fibres for sending data, and we store data optically on audio CDs and DVDs.

Using beams of light instead of electrons inside a computer gives you two main advantages. First, you can process information 1000 times faster using very little power. Second, you can do huge numbers of connections and calculations, all at the same time. We're a long way from building an all-optical computer but people are working hard to produce one.

DNA Computers

Yes, DNA computing uses DNA, the stuff in your genes.

In November 1994, Dr Leonard Adelman of the University of Southern California shocked the world of computer science when he wrote a paper describing how he had used a tiny drop of DNA molecules to solve the famous Travelling Salesman Problem. Adelman was already famous through a very well-known method of encrypting data called the RSA method. Adelman is the 'A' of RSA!

The Travelling Salesman Problem is a few centuries old. You want your salesman to visit each city only once and not double back on his path. It's easy to solve for three cities. But by the time you get to 100 cities, you need massive computing power just to get a close-enough solution to this simple problem.

If you suspend 500 g of DNA molecules in one tonne of water, you have more storage capacity than all the computers built in the history of the human race. (DNA molecules are very small, so you need lots of them to make up 500 g.) I won't go into the details, but Adelman got tiny chunks of DNA to represent every possible route between all the cities. Then he basically filtered out what he didn't want.

If you think DNA computers are way out there, what about quantum computers?

Quantum Mechanics 101

Quantum computers can solve really enormous problems, because they don't just use the quantum computer on your desk. They use, at the same time, every other computer that could possibly be built in the past or future history of the Universe, anywhere in the Universe.

Quantum computers are really weird, because Quantum Mechanics is really weird. Quantum Mechanics is what happens in the world of the minutely small, where the atoms live.

The conventional image of an atom has electrons orbiting around a central core or nucleus. You would think that any given electron would just be orbiting around this central nucleus. It is, but this electron is also everywhere else in the entire Universe at the same time!

For example, think about something normal, like a rock or a stone. You know exactly where it is and if you leave it somewhere it will stay there. You can pick it up because you know where it is. You can change the shape of the rock with a chisel. You can even throw the rock accurately to anywhere you wish within throwing distance.

But if you think 'quantum', accuracy goes out the window.

Quantum Mechanics is Weird

If you try to look at a quantum particle, the mere act of observing it changes it into something else. And don't even think about trying to touch it.

You can't make your quantum particle have a specific shape, except by getting it to run between a few wave guides. And even in this case, you will get a fuzzy shape, not the exact one you wanted.

You can't 'store' a quantum particle — it will just evaporate. You also have problems with trying to make your quantum particle go from 'here' to 'there'. This is because at the same time as it is 'here' or 'there', it is also 'everywhere else in the Universe at the same time'.

In the computer logic that we use today, a normal memory element is either '1' or '0'. But a quantum memory element is truly weird. It is both '1' and '0' at the same time, as well as being in all the possible states in between '1' and '0'. This 'fuzziness' — its ability to be in all possible states at the same time — is what makes a quantum computer so powerful. When you use a quantum computer, you are also using a whole bunch of other quantum computers in other universes at the same time.

Quantum Computers

We haven't actually built a quantum computer yet, but we have proved that it is not impossible. And almost every month, another scientist somewhere in the world makes another advance that brings us a little closer to a working quantum computer.

The great attraction of quantum computers is the immense power that they will have.

Consider credit card security. Think about a number that has 100 digits in it. This is a really big number. The number which tells you how many atoms there are in the entire Universe has only 80 digits in it! Now try to work out the only two numbers which, when multiplied together, will give you the original 100-digit number. This is called 'factoring a 100-digit number'. It's a real and useful problem. Most computer security systems use exactly this logic to protect credit card numbers.

It is very easy to multiply two big numbers together to get an even bigger number. But going backwards is very difficult. There is no mathematical theory on how to do it. You just have to use the brute force method, i.e. try every possible solution. To factor a 250-digit number would take the best part of 1000 years. But a quantum computer could factor a 250-digit number in seconds!

That's why quantum computers are so attractive.

Faster Computers?

What on earth would we do with this super computing capacity?

For one thing, we would be able to get much better weather forecasts around the world. For another, we would be able to model the action of a drug even before we put it together. Today it takes about 10 years and half-a-billion dollars to put a new drug on the market. If the drug shows some nasty side effects anywhere along this pathway, it gets thrown out, and all the effort and money are wasted.

In the year 2050, a computer could be something very different, smaller than a grain of sand perhaps, and getting its power from its environment. It could be painted onto a bridge, making the structure stronger and safer. It could be your auxiliary brain, storing all the information in all the databases of the world and implanted in your biological brain. It could help you speak every language on the planet.

If Moore's Law does run to its ultimate conclusion, technology should be able to give us incredibly small and very powerful computers. It could mean fewer working hours, less environmental devastation, less greed and poverty and less war and suffering.

Or it could mean Moore of Less for all of us?

Plenty of Room at the Bottom

Richard Feynman, the Nobel Prize–winning physicist, once gave a lecture called 'There's Plenty of Room at the Bottom'.

He claimed that if we were ever able to write one piece of information on an atom, we could store every bit of information ever generated by the entire human race in something smaller than a grain of sand. This includes every book ever written, every thought, every painting, every gesture — everything any human being has ever done.

The Big Con

In the science fiction stories that I read in my youth, there was one underlying message — computers and robots would give us so much leisure time that we would need to learn how to spend it.

Today, people with jobs are all working at least 10% longer hours than they used to. What went wrong?

References

Muller, D. A., et al., 'The electronic structure at the atomic scale of ultrathin gate oxides', *Nature*, 24 June 1999, pp 758-761.

Negroponte, Nicholas, 'Hack out the useless extras', *New Scientist*, 5 June 2004, p 26.

Reed, Mark A., et al., 'Computing with molecules', *Scientific American*, June 2000, pp 68-75.

Schultz, Max, 'The end of the road for silicon?', *Nature*, 24 June 1999, pp 729, 730.

Service, Robert F., 'Can chip devices keep shrinking', *Science*, 13 December 1996, pp 1834-1836.

'Steering the future of computing', *Nature*, 23 March 2006, p 383.

Acknowledgments

Gone are the days when a person could write a book by themselves — including this section ...

With this in mind, I would like to thank the following people:

Mary Lou (I love you — first and final edits), Dan 'The Man' Driscoll (ABC Star Producer), all the wonderful editorial staff at the *Good Weekend*, my fabulous agents, the 'Lovely' Lesley McFadzean and the 'Super' Sophie Hamley for their advice and expertise on all things literary, the very artistic Adam 'Scissors' Yazxhi, my excellent editor Lydia Papandrea, the Amazing Anabel Pandiella for the publicity angles, my perfect publishers Alison 'The Axe' Urquhart and Shona Martyn, the legend that is Judi Rowe, Jill Donald for perfect production and patience, the family — Carmel, Max, Lola, Alice and youngun' Karl — for story suggestions and Caroline P. eggers Pegram for stuff in general.

Hitched and Honeymoon Triumphal Tour 2006

London Eye

Before we left

Wedding bouquet: 3 Freesias for $75

Church in the Arctic Circle

Inside the Church

Just married

Wedding reception

Reindeer

Midnight sun

Sunnies and champagne at midnight

Norwegian fjord

Norway

Salomea Kruszelnicka's grave

Prague